What is Anthropology?

The word "anthropology" derives from the Greek and literally means "the study of man" or "the science of man". But the "man" of anthropology was a special kind of "man".

HISTORICALLY, ANTHROPOLOGY WAS THE "STUDY OF PRIMITIVE MAN".

I AM ANAZASI. THEY CALLED ME A PRIMITIVE MAN.

What is "Primitive"?

In *The Mind of Primitive Man* (1938), **Franz Boas** (1858–1942), founder of American Cultural Anthropology, told us just who are the primitives.

INTRODUCING

Anthropology

Merryl Wyn Davies • Piero

Edited by Richard Appignanesi

Icon Books UK ◆ Totem Books USA

This edition published in the UK in 2005 by Icon Books Ltd., The Old Dairy, Brook Road, Thriplow, Cambridge SG8 7RG email: info@iconbooks.co.uk www.iconbooks.co.uk

This edition published in the USA in 2005 by Totem Books Inquiries to: Icon Books Ltd., The Old Dairy, Brook Road, Thriplow, Cambridge SG8 7RG, UK

Sold in the UK, Europe, South Africa and Asia by Faber and Faber Ltd., 3 Queen Square, London WC1N 3AU or their agents

Distributed to the trade in the USA by National Book Network Inc., 4720 Boston Way, Lanham, Maryland 20706

Distributed in the UK, Europe, South Africa and Asia by TBS Ltd., Frating Distribution Centre, Colchester Road, Frating Green, Colchester CO7 7DW

Distributed in Canada by Penguin Books Canada, 90 Eglinton Avenue East, Suite 700, Toronto, Ontario M4P 2YE

This edition published in Australia in 2005 by Allen and Unwin Pty. Ltd., PO Box 8500, 83 Alexander Street, Crows Nest, NSW 2065

ISBN 1 84046 663 4

Previously published in the UK and Australia in 2002

Text copyright © 2002 Merryl Wyn Davies
Illustrations copyright © 2002 Piero

Originating editor: Richard Appignanesi

Printed and bound in Singapore
by Tien Wah Press Ltd.

Studying People

Anthropologists study people. They study how people live, human society past and present. Anthropology is also about how we think about people thinking about people, now and in history. And sometimes it is about power relations between people, peoples, cultures and societies, colonialism and globalization.

ANTHROPOLOGY IS ...

- The study of man from biological, cultural and social viewpoints.

- The study of human cultural difference.

- The search for generalizations about human culture and human nature.

- The comparative analysis of similarities and differences between cultures.

Anthropology's Big Problem

The biggest problem in anthropology is how to talk about its object of study. *Primitive*, *savage* and *simple* are prejudicial, discriminating and supremacist terms. Yet they defined the people anthropologists were particularly interested to study and why they wanted to study them.

THE FUNDAMENTAL SPIRIT OF ANTHROPOLOGICAL RESEARCH CONSISTS IN THE APPRECIATION OF THE NECESSITY OF STUDYING **ALL FORMS OF HUMAN CULTURE**, BECAUSE THE VARIETY OF ITS FORMS CAN ALONE THROW LIGHT UPON THE HISTORY OF ITS DEVELOPMENT, PAST AND FUTURE.

WHAT ANTHROPOLOGISTS HAVE LEARNT AND ANTHROPOLOGY TRIES TO TEACH IS **WHAT IS WRONG** WITH THINKING ABOUT REAL PEOPLE AS PRIMITIVE, SAVAGE AND SIMPLE.

The Other

Today, anthropology is defined as the systematic study of the Other, while all other social sciences are in some sense the study of the Self. But, who is the Other and who the Self?

In *Reinventing Anthropology* (1969), **Dell Hymes** wrote: "the very existence of an autonomous discipline that specializes in the study of Others has always been somewhat problematic."

The Changing Problem

How anthropology deals with its "problem" is now a topic of heated debate internal to anthropology. And two other things have changed. First: the Other has changed. Non-Western societies have undergone rapid social change.

I REFUSE TO BE THE VANISHING NOBLE SAVAGE. I DEMAND MY RIGHTS AND THE RIGHT TO BE TREATED JUST LIKE YOU.

WE WANT MANHATTAN BACK!!

Second: anthropology has come home. It no longer exclusively studies non-Western cultures. Now anthropologists also study marginal cultures in Western societies as well as institutional and organizational cultures, such as business corporations, scientists and the police.

How does anthropology cope with these changes? It studies the history of anthropology itself, the assumptions of anthropologists past and present, the reactions of anthropologists past and present – and it ponders whether anthropology tells us more about the Self than the Other.

"First, it is hard to say what it is the study of; secondly, it is not at all clear what you have to do to study it; and thirdly, no-one seems to know how to tell the difference between studying anthropology and practising it."

Tim Ingold, Professor of Anthropology, University of Aberdeen

The Origins of Anthropology

"What makes anthropology anthropology is not a specific object of enquiry, but the history of anthropology as a discipline and practice."
Henrietta Moore, Professor of Social Anthropology, London School of Economics

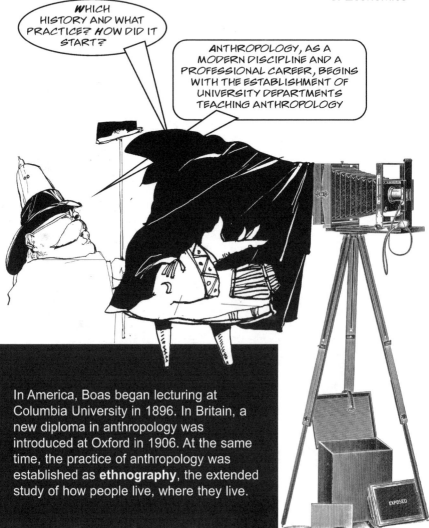

WHICH HISTORY AND WHAT PRACTICE? HOW DID IT START?

ANTHROPOLOGY, AS A MODERN DISCIPLINE AND A PROFESSIONAL CAREER, BEGINS WITH THE ESTABLISHMENT OF UNIVERSITY DEPARTMENTS TEACHING ANTHROPOLOGY

In America, Boas began lecturing at Columbia University in 1896. In Britain, a new diploma in anthropology was introduced at Oxford in 1906. At the same time, the practice of anthropology was established as **ethnography**, the extended study of how people live, where they live.

The Founding Fathers

Alan Barnard in *History and Theory of Anthropology* (2000) names the French philosopher **Charles Montesquieu** (1689–1755) as the common ancestor of all modern anthropology. Anthropology begins in 1748 with the publication of his *The Spirit of the Laws*. It is a product of the Enlightenment.

Bronislaw Malinowski

Lewis Henry Morgan

Charles Montesquieu

Sir
Edward
Burnett
Tylor

Sir Henry
Sumner Maine

Then comes the Darwinian horizon in the 1860s when great names – **Sir Henry Sumner Maine** (1822–88), **Lewis Henry Morgan** (1818–81), **Sir Edward Burnett Tylor** (1832–1917) and **Sir James Frazer** (1854–1941) – define the intellectual tradition that leads to modern anthropology. In 1871, the Anthropological Institute is founded in London. When Franz Boas, **Bronislaw Malinowski** (1884–1942) and **A.R. Radcliffe-Brown** (1881–1955) establish the practice of ethnography, modern anthropology is underway.

The Hidden Agenda

Marvin Harris in *The Rise of Anthropological Theory* (1968) also argues for the Enlightenment origins of anthropology. Many more Enlightenment men, including **Denis Diderot** (1713–84), **Jacques Turgot** (1727–81) and **Marquis de Condorcet** (1743–94), enter the list.

Harris also gives a nod towards the French writer, **Michel de Montaigne** (1533–92), whose essay "On the Cannibals" was published in 1580.

Michel de Montaigne

BUT IN A BRACKETED ASIDE AND FOOTNOTE, HARRIS DISMISSES AS IMPERTINENT THE SUGGESTION OF ANY REAL CONNECTION BETWEEN *ENLIGHTENMENT* IDEAS AND WHAT WENT BEFORE.

WELL, HE WOULD, WOULDN'T HE? *T*ELL THEM WHY ...

The Age of Reconnaissance

Montaigne met some South American Indians brought to perform at a fair in France. He then wrote his classic essay that framed non-Western peoples by their lack of the defining attributes of civilization.

Montaigne's ideas were formed not by "experience" but speculation. His speculation is part of a vast literature about the "new" people discovered since Christopher Columbus's pratfall in America and Vasco da Gama being guided to India in what is called the "Age of Reconnaissance" …

THIS WAS THE PERIOD WHEN *EUROPEANS* DRAMATICALLY EXPANDED THEIR GEOGRAPHICAL HORIZONS AND KNOWLEDGE.

AND SLAUGHTERED, ENSLAVED AND DECIMATED. *W*ITHOUT ME AND MY HISTORY, THERE'S NO ANTHROPOLOGY, AND THAT'S WHAT THEY DON'T WANT TO ADMIT!

"Fidelity to the Old"

What Harris, tartly, wants to denounce and exclude is the argument of Margaret Hodgen. In her *Early Anthropology in the Sixteenth and Seventeenth Centuries* (1964), Hodgen made two telling points.

First, speculation about human origins, ways of life and diversity is old, interactive and continuous. The concepts and ideas of the Ancient Greeks, medieval writers, the Age of Reconnaissance, Montaigne and much more, inform and construct Enlightenment ideas and the intellectual tradition of 19th-century anthropology.

SECOND, THE ORGANIZING PRINCIPLES AND THEORETICAL IDEAS OF THESE ARCHAIC LAYERS OF SPECULATION KEEP ON RECURRING AND ARE ALIVE AND WELL IN **MODERN** ANTHROPOLOGY.

THIS IS WHAT *HODGEN* CALLS THE "MIND'S FIDELITY TO THE OLD WHICH HAS LEFT ITS MARK ON ANTHROPOLOGY."

What characterizes the early writing Hodgen links to anthropology? One strand is belief in the "Plinian People", so called from the section in the Roman writer Pliny the Elder's *Natural History* (AD 77) recording a large collection of monstrous races – dog-headed people, people with no heads and *anthropophagy* (cannibalism) – living at the fringes of the known world. These monstrous races were a standard feature of classical and medieval writings. Another strand is the biblical framework of explanation.

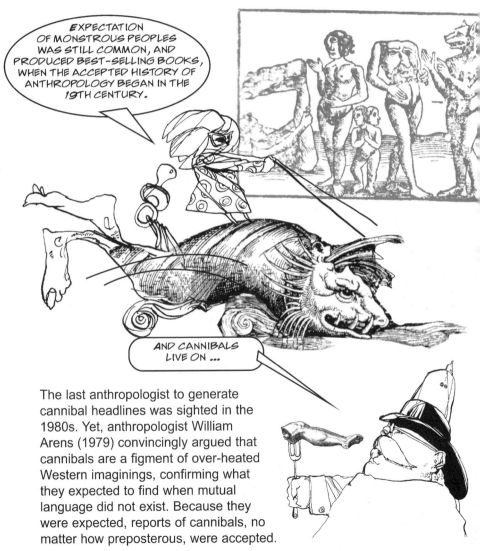

EXPECTATION OF MONSTROUS PEOPLES WAS STILL COMMON, AND PRODUCED BEST-SELLING BOOKS, WHEN THE ACCEPTED HISTORY OF ANTHROPOLOGY BEGAN IN THE 19TH CENTURY.

AND CANNIBALS LIVE ON ...

The last anthropologist to generate cannibal headlines was sighted in the 1980s. Yet, anthropologist William Arens (1979) convincingly argued that cannibals are a figment of over-heated Western imaginings, confirming what they expected to find when mutual language did not exist. Because they were expected, reports of cannibals, no matter how preposterous, were accepted.

The Question of Human Rights

Anthony Padgen, the Cambridge historian who specializes in the study of Spanish thinking about the "New World", provides similar arguments.

He makes a first important point. The public debate by the Catholic Church about the human or non-human status of American Indians held at Valladolid, Spain, in 1550 – echoing on till the 1570s – sets the parameters in which anthropological thought and argument operate.

THE CASE FOR THE PROSECUTION CAME FROM JUAN GINÉS DE SEPÚLVEDA, CHAPLAIN AND OFFICIAL CHRONICLER OF THE KING OF SPAIN.

DRAWING ON THE GREEK PHILOSOPHER ARISTOTLE, I PRESENTED NON-WESTERN PEOPLES AS EITHER NATURAL BARBARIANS OR NATURAL SLAVES AND CHILDREN.

Bartolomé de Las Casas (1474–1566), a Dominican cleric, presented the opposing case: *"Defence Against the Persecutors and Slanderers of the Peoples of the New World Discovered Across the Seas".* Las Casas knew what he was talking about.

I'D BEEN IN THE *AMERICAS* SINCE *1502.* I'D OWNED *INDIAN* SLAVES AND PROFITED FROM THEIR EXPLOITATION - AS REVEALED IN MY *HISTORY OF THE INDIES* (*1566*).

AFTER *1515*, HE DEDICATED HIS LIFE TO THE ADVOCACY OF *INDIAN* RIGHTS. *W*HICH IS MORE THAN MOST PROFESSIONAL ANTHROPOLOGISTS HAVE DONE, EVEN AFTER THEY STARTED TALKING ABOUT SUCH THINGS IN THE *1950*s AND *60*s.

The Jesuit Relations

The real origin of anthropological fieldwork, argues Padgen, is not the accepted ancestors Boas and Malinowski but generations of Jesuit missionaries, especially those working in Canada: **Paul Le Jeune** (1634), **Jacques Marquette** (1673) and especially **Joseph Lafitau** (1724). The reports of their work were published in the annual *Jesuit Relations.*

THE *RELATIONS* PRESENTED INFORMATION ACQUIRED FROM LONG CONTACT AND INVOLVEMENT WITH INDIGENOUS PEOPLE ...

... AS WELL AS CONSIDERING WHAT THIS INFORMATION MEANT FOR THE UNDERSTANDING OF **HUMAN NATURE** - THE GENERALIZATION AND COMPARISON THAT ANTHROPOLOGY IS ALL ABOUT.

Jesuit Relations

CONQUISTADORES AND MISSIONARIES - NO WONDER THEY WANT TO KEEP THOSE ORIGINS QUIET!

Major Trends in Western Thought

An excellent corrective to the conventional history of anthropology as a discipline is provided by the anthropologist **William Y. Adams** (1927–) who echoes both Hodgen and Padgen. Adams looks at "major trends" in Western thought, ideas that operate "below the level of conscious theory".

> *THESE MAJOR TRENDS EXPLAIN WHERE ANTHROPOLOGY CAME FROM AND WHAT IT'S ABOUT ...*

Progressivism: the identification of human cultural history with progress on an upward escalator from "nasty and brutish" to the modern West that is always on top whenever you pick up the story.

Primitivism: the reverse idea, including nostalgia for primitive simplicity and the idea of degeneration, humankind marching downhill from the beginning, though some are saved by civilization.

Natural law: not recurrent behaviour but codes, behavioural prescriptions and restrictions common to all peoples and part of nature's (i.e., biological in origin) or God's (i.e., moral and cultural in origin) plan.

German idealism: based on the dualistic separation of mind (the substance of history) and matter (the substance of nature).

"Indianology": both popular ideology about American Indians – especially of the Noble Savage variety – and a major field of study centring on the otherness of the Other.

The Continuity of Tradition

These major philosophical trends are continuous traditions uniting the intellectual history of the West. They provide links between the ancient sources, through medieval times, Renaissance, Enlightenment and Victorian speculation, right through to the modern and even the postmodern era.

The Derived Minor Trends

Adams also identifies "minor trends" – not lesser but derivatives and applications of the major trends.

Rationalism: the belief in an ordered universe governed by laws that conform with and are comprehensible by human reason.

Positivism: broadly a label for **Empiricism**, a methodology comprising observation plus induction or deduction.

Marxism or **dialectical materialism:** a self-proclaiming ideology and "sect", clearly part of progressivism. Marx and Engels based their thinking on the work of American anthropologist **Lewis Henry Morgan** (1818–81), best known for his study of Iroquois Indians.

Utilitarianism and **socialism:** uniquely British schools of radicalism and approach to social reform less interested in the past and more focused on the future.

Structuralism: the belief in a structured universe or inherent and coherent structuring in nature's order that is not imposed by the observer; the structures are therefore universal. It is a derivative of natural law.

Nationalism: the predominant Western ideology of the last three centuries works to shape national traditions of anthropology and other social sciences.

AND LAST BUT NOT LEAST - THE BIG ENCHILADA! ...

Imperialism

Imperialism is best described as a practical strategy for exploitation, a field practice rather than a theory, an ideological framework in which Western thought operates. For anthropologists, subject peoples of the colonies were uniquely "theirs" to study. British anthropologist **Ernest Gellner** (1925–95) described the colonies as the "reserved laboratory" where anthropology took up its studies.

IN AMERICA, THE LABORATORIES WERE NEAR AT HAND ...

... THE RESERVATIONS WHERE NATIVE AMERICANS HAVE BEEN CONVENIENTLY CONFINED FOR THE SUMMER VISITS OF THE ANTHROPOLOGIST.

Ernest Gellner

The Complicity of Anthropology

Anthropologists trained colonial officials and reported to them –
although officials complained they never told them anything useful.

In his classic text, *Anthropology and the Colonial Encounter* (1973),
Talal Asad argued that anthropology served as the "handmaiden of
colonialism". The ideology of imperialism feeds on the same
intellectual and philosophic roots that produced anthropology,
making them partner projects. Anthropology did not create
colonialism, but its origins are certainly an epiphenomenon of
colonialism.

Violations of Ethics

In more recent times, anthropologists have collected detailed information in the service of US neo-imperialist conflicts in South-East Asia, a scandal that provoked ethical review within anthropology.

Back to the Roots

We have now seen the philosophical roots feed the conventional history of anthropology. These are the foundations on which it stands and shape how it operates. To link the roots to the modern anthropological tree and the four branches of the academic discipline, we need …

One character: "the primitive"

Two theories: Evolutionism and Diffusionism

One ideology: Race

The Indispensable Primitive

The intellectual tradition of anthropology takes shape in the 1860s with what British anthropologist **Adam Kuper** calls the *Invention of the Primitive Society: Transformations of an Illusion* (1988). "Primitive man" is the legitimate and defining object of study for which professional anthropology was created.

QUITE SIMPLY, ANTHROPOLOGY WITHOUT THE PRIMITIVE IS INCOMPREHENSIBLE ...

*T*ODAY THEY DON'T KNOW WHAT TO CALL ME, CAN'T DO WITHOUT ME AND TIE THEMSELVES IN KNOTS TRYING TO SQUARE THE CIRCLE!

Speculating on the Invention

Invention is a conscious, constructive act. It begins with **Sir Henry Sumner Maine** (1822–88) who defined the distinction between primitive and civilized, as ascribed and acquired status.

SOCIETY ORIGINATES NOT IN A "SOCIAL CONTRACT" BUT IN THE FAMILY AND KINSHIP GROUPS BUILT ON THE FAMILY.

Ideas of kinship origins and the primitive condition were then speculated on and developed by **John Lubbock** (1834–1913), neighbour, friend and supporter of **Charles Darwin** (1809–82); the Scottish lawyer **John McLennan** (1827–81); the American **Lewis Henry Morgan** (1818–81) – all of whom were politicians as well as armchair scholars – and the Swiss **J.J. Bachofen** (1815–87).

What Came First?

McLennan and Morgan were arch–opponents. They argued vigorously.

Living Relics

Primal promiscuity, the precedence of *matriliny* over *patriliny*, and *matriarchy* over *patriarchy*, defined the primitive. These traits could be read in the living kinship practices of contemporary primitive societies. From this speculative construct, both the character of the primitive – its identification with *living peoples* – and the specialist field of anthropology begin.

The concept of the primitive was expanded by **Sir Edward Burnett Tylor** (1832–1917).

MY COLLECTING AND CLASSIFYING ACTIVITIES WERE FOUNDED ON THE IDEA OF "SURVIVALS" AND "RELICS".

THIS WAS AN OLD IDEA, FIRST SURFACING IN THE *1580S* WHEN AMERICAN *INDIANS* WERE USED AS MODELS TO EXPLAIN THE ANCIENT BRITONS.

Tylor too saw living "primitives" as relics of earlier stages of human existence.

Seen from the Armchair

Sir James Frazer (1854–1941), author of *The Golden Bough: A Study in Magic and Religion* (1890; enlarged 1907–15), established the mental, religious, magical and mythical components of the primitive.

Theories of Evolutionism

Social thought in the Western intellectual tradition had always been evolutionist. The hierarchy of stages of social evolution is derived from the Greek idea of the three ages: gold, bronze and iron.

The Enlightenment tradition, especially in the work of Scottish writer **Adam Ferguson** (1723–1816), identified these stages with the social and political formations of societies both living and dead, to make the threefold hierarchy of *savagery*, *barbarism* and *civilization*. The character of the primitive was merely fitted into this well-accepted scheme.

Integrating the Biological and Social

Darwin did not invent evolutionism. He introduced a particular theory of *biological* evolution. This quickly became the dominant paradigm, conflating ideas of the biological with the social, as it generalized the power of the concept evolutionism.

Herbert Spencer (1820–1903), a devout social evolutionist, had called for a biological theory of man before Darwin. Spencer is a proper anthropological ancestor …

All of modern anthropology is evolutionist in one sense or another. *Generalization*, *classification* and *typology* all implicitly or explicitly invoke evolutionary ideas and hierarchical relations.

The Theory of Diffusionism

Diffusion is the transmission of things from one culture, people or place to another. The essence of diffusion is *contact* and *interaction*. It is a very old idea. The biblical framework of explanation, developed vigorously in the 16th and 17th centuries, is diffusionist.

THE DISPERSION OF MANKIND WITH THE *FALL* OF THE *TOWER* OF BABEL PROVIDES A GENEALOGICAL - THAT IS, BIOLOGICAL AND SOCIAL DIFFUSIONARY - CONNECTION BETWEEN ALL PEOPLES.

DIFFUSION ALSO UNDERPINS THE DEVELOPMENT OF LANGUAGE STUDIES ...

Especially in the work of Orientalist **Sir William Jones** (1746–94) on the Indo-European language group and of the German **Max Müller** (1823–1900) who settled in Britain. Müller not only worked on comparative philology but was a proponent of the idea that all humanity shares the same mentality. Linguistics has been an important influence on social theory and especially anthropology.

Diffusionism is also the major theme in the work of other German scholars. **Friedrich Ratzel** (1844–1904) was the first to divide the world into culture areas. His work influenced Tylor. The work of **Leo Frobenius** (1873–1938) and **Fritz Graebner** (1877–1934) on the concept of *Kulturkreise*, or "culture circles", was an important influence on Boas.

In Britain, high diffusionism is the formula used by **Sir Grafton Elliot Smith** (1871–1937) and **William Perry** (1887–1949) in their book *Children of the Sun* (1923). They put forward the theory that the sun-worshipping Ancient Egyptians were the origin of all civilization. Smith and Perry substituted Egypt for the biblical and Hebrew patriarchs as the cradle of originality. They advanced diffusion as a rearguard action against evolutionism.

The Race Swindle

The primitive was indistinguishable from other races. Other races defined the concept of the primitive and provided the means to study the "primitive condition". Civilization was a singular term for the unitary medium through which the white race had risen to prominence, while all other races were left stalled at earlier levels of the evolutionary and racist hierarchy of savagery and barbarism.

Modern anthropology begins in contention with the concept of race. It denounces the racism in its intellectual origins and discounts the overwhelming mass of racist writing that informed these origins.

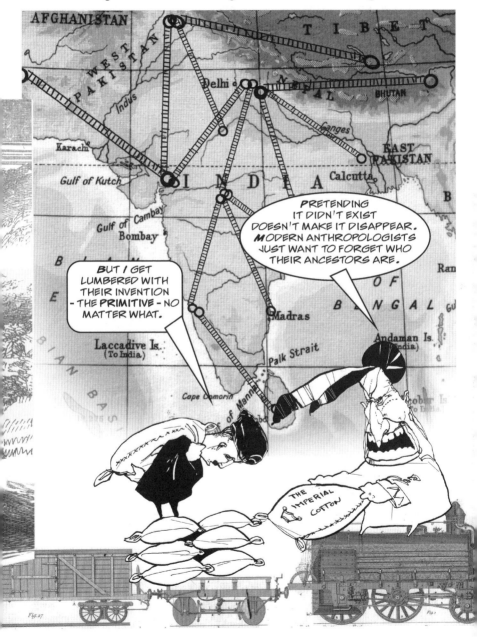

Field Studies

The dividing line between the historical era and professional modern anthropology is the shift from speculation to empirical science.

THE REAL MEETING-GROUND OF ALL BRANCHES OF ANTHROPOLOGY IS THE SCIENTIFIC STUDY OF CULTURE.

Bronislaw Malinowski (1884–1942)

It is signalled by a change of location: leaving the armchair, the philosopher's ivory tower and the colonial veranda to venture off into the field where "primitive" people lived.

In the field, empirical inquiry would provide new empirical evidence. This would explain different cultures and permit comparison of human similarities and differences that was not pure speculation.

THAT'S WHAT THEY WANT US TO BELIEVE – EXCEPT FOR ONE THING. *EXPLAIN*, PLEASE!

The theories and organizing principles that accompanied the anthropologist into the field were those inherited from his ancestors – both acknowledged and unacknowledged.

The Anthropological Tree

There are four branches of study that make up anthropology.

The signpost reads:
PHYSICAL OR BIOLOGICAL ANTHROPOLOGY
ARCHAEOLOGY
ANTHROPOLOGICAL LINGUISTICS
SOCIAL OR CULTURAL ANTHROPOLOGY (aka Ethnology)

MOST UNIVERSITIES IN THE *U*NITED *S*TATES INTRODUCE STUDENTS TO ALL FOUR BRANCHES. *FEW BR*ITISH UNIVERSITIES *DO*, CONCENTRATING INSTEAD ON SOCIAL ANTHROPOLOGY.

The vast majority of anthropologists work in social anthropology. In the US, they call it cultural anthropology or even ethnology – which is what British anthropologists used to call their subject before it became a university discipline.

The naming of parts – as either *social* or *cultural* anthropology – is a historic legacy of the specific development of the British and American national traditions. Various nationalities have their own differences of emphasis and naming in their anthropologies.

Physical Anthropology

Physical anthropology began as the study of the races of man. *Anthropometrists* with their callipers set about their favourite occupation of measuring and classifying head sizes.

The objective was to prove racial differences as physically given and support racist theories of both human origins and cultural diversity.

Polygenesis vs. Monogenesis

The big debate in physical anthropology was between the **polygenists** and the **monogenists**. **Polygenesis** means the origin of different races (or species) from different stocks. It was an early idea developed to explain the American Indians, first speculated about by **Isaac de la Préyère** (1594–1676).

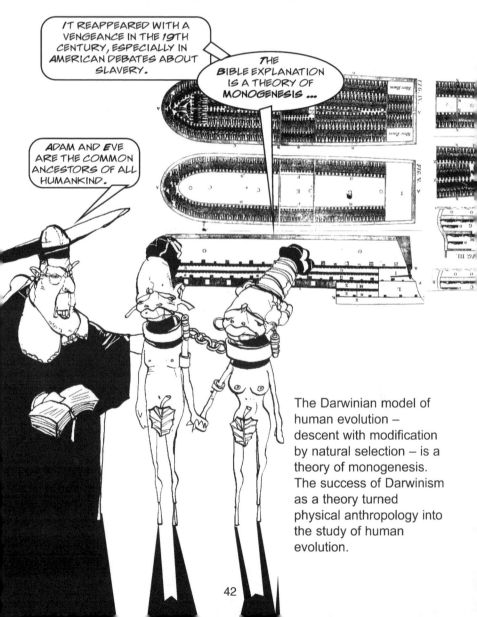

IT REAPPEARED WITH A VENGEANCE IN THE 19TH CENTURY, ESPECIALLY IN AMERICAN DEBATES ABOUT SLAVERY.

THE BIBLE EXPLANATION IS A THEORY OF MONOGENESIS ...

ADAM AND EVE ARE THE COMMON ANCESTORS OF ALL HUMANKIND.

The Darwinian model of human evolution – descent with modification by natural selection – is a theory of monogenesis. The success of Darwinism as a theory turned physical anthropology into the study of human evolution.

Human Ecology and Genetics

Physical anthropology includes studying classification – from tooth variations between monkeys and modern man to comparative anatomy and physiology. Both human ecology and genetics are part of physical anthropology.

HUMAN ECOLOGY STUDIES THE ADAPTIVE RESPONSES OF HOMO SAPIENS IN DIFFERENT ENVIRONMENTAL CONDITIONS ...

... AS WELL AS THE ECOLOGY OF DISEASE, NUTRITION AND DEMOGRAPHY.

Genetics in anthropology concerns the genetic variation of different racial groups. It is overshadowed by the growth of biological genetics.

The Rise of Sociobiology

Physical anthropology fell out of fashion both because of its racist associations and because it was overtaken by the rise of modern biological sciences. But it came back with a vengeance in the 1970s and 80s through the development of **sociobiology** – the study of the genetic basis of human behaviour.

THE RELEVANCE AND MEANING OF *SOCIOBIOLOGY* IN ANTHROPOLOGY ARE THE SUBJECTS OF A MAJOR CONTEMPORARY DEBATE.

IT'S WHERE ALL THE UNWANTED HISTORY COMES HOME TO ROOST.

A Refocus of Race in Gene Theory

Gene-centred theory recaptures a great deal of the 19th-century idea of the primitive, with *behaviour-determining genes* operating as a new reformulation of race. The models for early human behaviour now switch to *animal* behaviour.

*T*HE FOCUS OF INTEREST IS THE "ENVIRONMENT OF EVOLUTIONARY ADAPTATION" (*EEA*) WHICH OCCURS ON THE AFRICAN SAVANNAH.

*T*HE "OUT OF AFRICA" THESIS IS A STUDY OF HUMAN EVOLUTION WITH STRONG OVERTONES OF DIFFUSION.

Other Links with Early Anthropology

Genes are studied in populations. The essence of the population group is *interbreeding* and *control of reproduction*. This is exactly why family and kinship were originally used to construct the concept of "primitive society" and the character of the primitive.

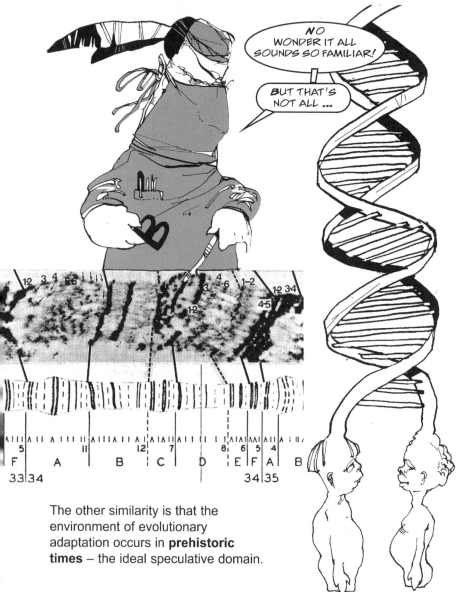

The other similarity is that the environment of evolutionary adaptation occurs in **prehistoric times** – the ideal speculative domain.

Archaeology and Material Culture

Archaeology and anthropology share a common interest in seeking to explain the origins of culture and society, and the development of civilization. *Material culture* is the study of the techniques of producing the material goods and means of production of societies by anthropologists. It is the study of everything from the techniques of making pottery to 50 ways to castrate a camel.

Anthropological Linguistics

Linguistics and anthropology had the same kind of relationship as anthropology and archaeology for most of the 19th and 20th centuries – that is, a shared interest in the study of exotic languages to trace links between them and their historical development.

LINGUISTICS THEN UNDERWENT A MAJOR REVOLUTION WITH THE RISE OF TRANSFORMATIONAL AND GENERATIVE THEORIES.

NOAM CHOMSKY (1928–) IN PARTICULAR AIMED TO UNCOVER THE UNDERLYING PRINCIPLES OF ALL LANGUAGES – A "UNIVERSAL GRAMMAR".

Linguistic concepts and theories are borrowed by anthropologists. Linguistic models are used as models for cultural and social behaviour by structuralist and cognitive social anthropologists who see societies as *communicative systems* and who regard language as the basis for modes of thought.

Social or Cultural Anthropology

Named either "social" or "cultural" anthropology, this is the principal branch of the discipline in which grand theory – or theorization of any kind – is developed. It involves the study of cultural diversity, the search for cultural universals, the study of societies as functioning wholes, the study of social structure, the interpretation of symbolism, and much else.

What is Culture?

The principal distinction between American cultural and British social anthropology is the difference between a focus on culture as the "whole" that the anthropologist is studying, and society, its structure and organization, as the "whole" within which culture operates. On both sides of the Atlantic, there are myriad definitions of culture. In 1952, the leading American anthropologists **A.L. Kroeber** (1876–1960) and **Clyde Kluckhohn** (1905–60) cited over 100.

YET THE MOST FAMILIAR TO ALL ANTHROPOLOGISTS IS THE CANONICAL DICTUM OF *E.B. TYLOR* IN *PRIMITIVE CULTURE* (1871).

CULTURE IS THAT COMPLEX WHOLE WHICH INCLUDES KNOWLEDGE, BELIEF, ART, LAW, MORALS, CUSTOMS, AND ANY OTHER CAPABILITIES AND HABITS ACQUIRED BY MAN AS A MEMBER OF SOCIETY.

For Tylor, culture was a singular term, the domain in which all human society developed in an evolutionary progression from simple to complex. Modern anthropology as a professional discipline begins from the idea of *cultures* – a plurality of ways of life that must be understood *in their own terms*.

TODAY, THE CONCEPT OF CULTURE HAS BECOME, AS *HENRIETTA MOORE* (1957-) PUTS IT ...

A SERIES OF SITES OF CONTESTED REPRESENTATION AND RESISTANCE WITHIN FIELDS OF POWER.

The American anthropologist Roy Wagner argues: "the core of culture is … a coherent flow of images and analogies that cannot be communicated directly from mind to mind but only elicited, adumbrated, depicted." Rather than being stable systems of collective representations, cultural meanings "live in a constant flux of continual re-creation".

Increasing Specializations

Tylor's shopping list of cultural traits still serves to introduce the fields of specialization in social or cultural anthropology: social organization, economic anthropology, political anthropology, anthropology of art, religion, law and kinship studies.

From applied, action, cognitive, critical and development anthropology, via feminist, Marxist and medical anthropology, through to symbolic and visual anthropology. These divisions become distinguished as sub-fields, topics and theoretical orientations.

The Bedrock of Ethnography

Literally the *writing* of culture, ethnography is the basic practice of all social and cultural anthropology, involving fieldwork and allegedly objective and scientific observation. Ethnography provides anthropology with its raw material, its major conceptual and method-ological claim to fame – "participant observation" – and its reason for being: the bedrock for comparison, generalization and theory.

Writing the Exotic

The ethnography of particular regions and their inhabitants is a sub-field or specialization within anthropology. Some prime examples are Melanesia, West Africa, Australian Aborigines and Amazonian Indians. Writing about "exotic" peoples shaped the language of anthropology. The form, content, questions and interests of ethnography chart the debates and changes in anthropology.

OLD ETHNOGRAPHIES DON'T JUST FADE AWAY. THEY RECUR AND FEED WHAT ANTHROPOLOGISTS THINK AND DO.

The importance of ethnography was emphasized by two of the biggest names in modern anthropology – Franz Boas and Bronislaw Malinowski.

Franz Boas

The founder of American anthropology, **Franz Boas** (1858–1942), was born in Minden in Germany and initially trained in physics and geography. In 1883, he joined an expedition to Baffin Island and began fieldwork with the Inuit (Eskimo).

THREE YEARS LATER, I BEGAN WORK IN BRITISH COLUMBIA AMONG THE KWAKIUTL.

In 1896, he joined Columbia University, New York, and became their first Professor of Anthropology in 1899, a position he held for 37 years. Boas trained most of the next generation of American anthropologists.

Bronislaw Malinowski

Bronislaw Malinowski (1884–1942) is considered to be the founder of British anthropology. He was born in Krakow, Poland, and studied mathematics and physics there before moving to study in England. His interest in anthropology was sparked by reading Sir James Frazer's *The Golden Bough*.

Returning to England, he took up a position at the London School of Economics and was appointed as their first Professor of Anthropology in 1927.

Malinowski trained many of the first generation of British anthropologists.

MALINOWSKI AND I AGREED ON THREE POINTS ...

An emphasis on participant observation

Long-term exposure to the society being studied

Use of the native language

BUT WHEREAS BOAS EMPHASIZES THE MINUTE DETAILS OF CULTURE, *I* STRESS THE FUNCTIONS OF **SOCIAL INSTITUTIONS** IN RELATION TO THE INDIVIDUAL.

Fieldwork

To produce ethnography, anthropologists do fieldwork. It is the **rite of passage** that makes an anthropologist. The oldest guide to fieldwork is *Notes and Queries in Anthropology* which first appeared in 1874. *Notes and Queries* was produced by the British Association for the Advancement of Science. The section on culture was written by E.B. Tylor.

Notes and Queries was revised and edited in 1951 by the Royal Anthropological Institute of Great Britain and Northern Ireland (RAI).

Human Ecology in Fieldwork

The first thing one needs to know when setting out on fieldwork is where a people live, what kind of environment they inhabit and what kind of subsistence base and economy they operate. So, one may come across …

Hunters and Gatherers
The Bushmen or San of South Africa; Pygmies of Central Africa; Hadsa in East Africa; Australian Aborigines; Andaman Islanders; Inuit (Eskimo); Algonkins (such as Cree) and other groups in Canada.

Fishing
This may be the basis of hunting and gathering societies, as among the Kwakiutl and other north-west-coast groups in Northern America.

Pastoralists or Herders
People who depend on their livestock: the Tuareg and Fulani in West Africa; Nuer and Masai and other groups in East Africa; Bedouin in the Middle East; Saami (or Lapps) in Northern Europe.

Settled Cultivators or Horticulturalists
Most of Africa, South and South-East Asia and New Guinea; most Amazonian peoples; Ojibwa and other north-eastern groups in Northern America; Hopi, Navaho, Pueblo and other groups in the American south-west; peasant communities in Southern Europe.

Understanding how a society exploits its environment means investigating the seasonal cycle, asking how the environment is understood, how work is divided between members of the community, and finding out what ritual and ceremonial beliefs and practices are involved in the business of making a living.

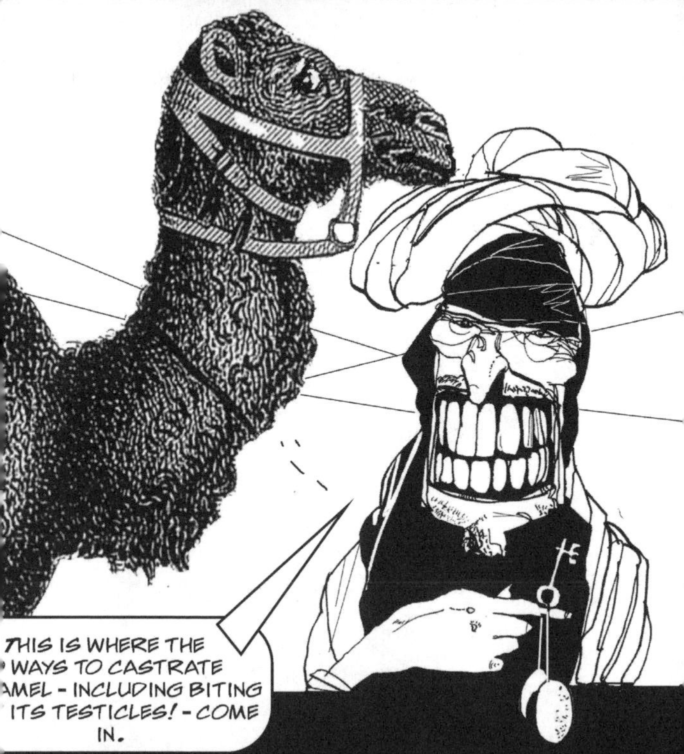

THIS IS WHERE THE
WAYS TO CASTRATE
AMEL – INCLUDING BITING
ITS TESTICLES! – COME
IN.

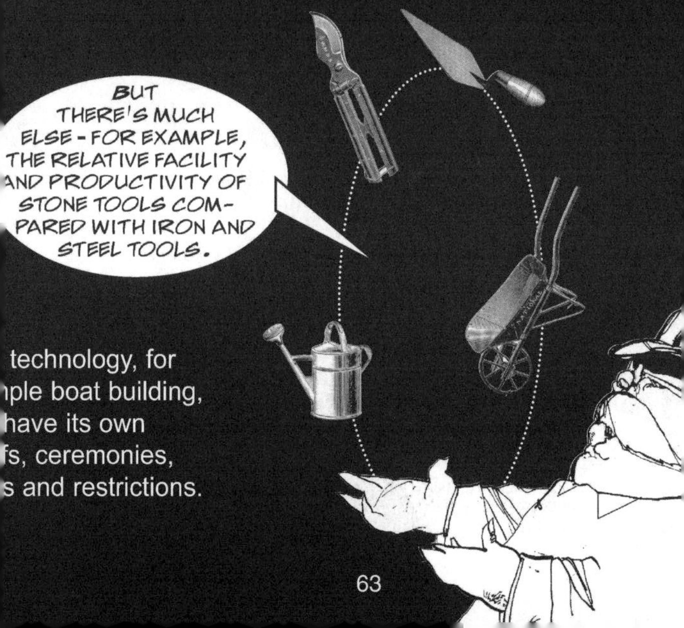

BUT
THERE'S MUCH
ELSE – FOR EXAMPLE,
THE RELATIVE FACILITY
AND PRODUCTIVITY OF
STONE TOOLS COM-
PARED WITH IRON AND
STEEL TOOLS.

technology, for
mple boat building,
have its own
s, ceremonies,
s and restrictions.

Ecological Anthropology

Julian H. Steward (1902–72) introduced ecological anthropology in his *Theory of Culture Change* (1955). He argued that environment and technology play a major role in determining the social organization of culture and can be correlated with an evolutionary framework.

THE MAIN CONCEPTS OF ECOLOGICAL ANTHROPOLOGY ARE ...

Julian H. Steward

Adaptation – the ability to respond to environmental stress.

Means of subsistence – the method of exploiting an environment, such as fishing, hunting, gathering, herding or agriculture.

Ecological niche – the set of resources used in a particular environment. Different people may exploit different ecological niches in the same environment.

Carrying capacity – the maximum number of people, following a particular means of subsistence, who can exist in a particular environment.

The Question of Economy

How much food and goods are produced leads to the questions of
how the economy is organized, whether a surplus is produced and
what happens to it. Access to and distribution of economic resources
may be determined on very different principles and be involved with
ritual and ceremonial relationships.

HAVING
ACCESS TO LAND
MAY DEPEND ON
KINSHIP, MEMBERSHIP
OF A FAMILY OR OF A
GROUPING OF
FAMILIES.

PRODUCTION OF
GOODS AND SERVICES MAY
ALSO BE BASED ON KINSHIP
OR BIRTH.

EXCHANGING
GOODS MAY BE RELATED
TO SEEKING POWER AND
INFLUENCE IN SOCIETY.

"Potlatch", "Big Men" and *kula* are three examples of the
way in which economic concepts work in pre-capitalist
societies.

The Potlatch Ceremony

Kwakiutl society in west-coast Canada produces a surplus, but this is used to hold great ceremonials in which surplus produced by one kin group is distributed to other kin groups. It equalizes the distribution of produce and confers prestige on those who give.

IT CREATES **RECIPROCAL** OBLIGATIONS WITH THOSE WHO RECEIVE ...

WE WILL BE OBLIGED TO HOLD A POTLATCH AT SOME FUTURE TIME.

The "potlatch" itself, after which the ceremony is named, is a brass plate which may be destroyed along with other ceremonial goods as part of the ceremony.

The "Big Men" of New Guinea

In New Guinea, surplus economic resources – especially pigs – are accumulated and then distributed as gifts.

The Kula Exchange

In the Trobriand Islands of the Western Pacific, islanders operate spheres of exchange in which arm bracelets are exchanged for necklaces. The exchanges are made by chiefs and other powerful men, and confer status each time an exchange is made.

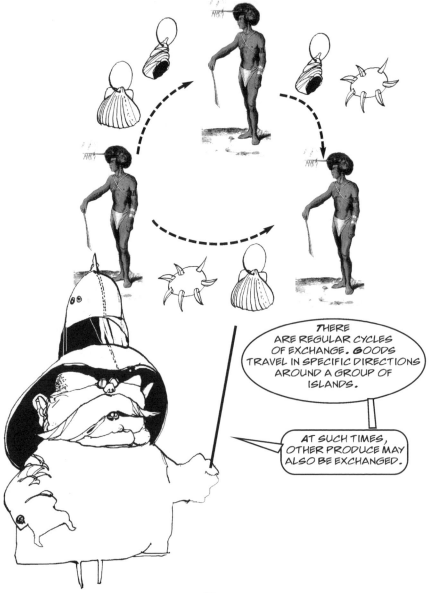

Economic Anthropology

The Gift (or *Essai sur le don*), published in 1925 by the French anthropologist **Marcel Mauss** (1872–1950), laid the foundation for economic anthropology. Mauss noted that a gift is never free; it entails three obligations: *to give*, *to receive* and *to reciprocate*.

RECIPROCITY MAY BE **IMMEDIATE** OR **DELAYED** – TO BE FULFILLED AT A LATER TIME.

Marcel Mauss

THE OBLIGATIONS INCURRED MAY BE MATERIAL OR MAY INVOLVE DEFERENCE, ACKNOWLEDGING THE SUPERIORITY OF THE GIVER.

Exchange and Trading Networks

Exchange may be made through markets and involve extensive trading networks. Women are often important market traders – for example, in West Africa and South-East Asia.

MONEY AS A MEDIUM OF EXCHANGE COMES IN MANY FORMS AND MAY MEAN MORE THAN MERE CURRENCY.

In her study of the Lele of Kasai in the Congo, **Mary Douglas** (1921–) showed that the raffia cloths woven by the Lele were used as a means of exchange, while neighbouring groups merely wove raffia for clothing. Lele raffia cloth performed four distinct functions …

- clothing
- used in formal gift-giving or in status payments among kin
- used as money, setting the value of goods exchanged between non-kin
- used in exchange to acquire a variety of goods, from hoes to pottery, from other peoples – when it was money to the Lele but barter to the recipients

The Formalist–Substantivist Debate

The great debate in economic anthropology is between the Formalists and Substantivists, and concerns the question of whether the "laws" of economics are universal.

Formalists argue that economics is a science and that economic anthropology is closely related. Economic rationality is a basic "law". People choose what is in their best interest and reject what is not.

THIS PERMITS COMPARISONS TO BE MADE BETWEEN RADICALLY DIFFERENT CULTURES.

Substantivists argue against the universal "law" of economics, especially the idea of "economic rationality". Instead, they emphasize economics as embedded in culture. Different spheres of exchange operate differently in various societies.

THERE ARE DIFFERENT ATTITUDES TO EXCHANGE AND WORK, AND DIFFERENT CONCEPTS OF VALUE ATTACHED TO THE SAME GOODS IN DIFFERENT SOCIETIES.

The economist **Karl Polanyi** (1886–1964) in *Trade and Markets in the Early Empires* (1957) made the distinction between the "substantive" meaning of the economic – the relation between making a living and one's natural and social environment – and the "formal" meaning of the economic – the logical relation between means and ends.

The classic collection of essays on both sides of the debate is *Economic Anthropology* (1968), edited by Edward E. LeClair and Harold K. Schneider.

Marxist Anthropology

Both Formalist and Substantivist positions come under criticism from Marxist anthropology. However, Marxist anthropology itself includes aspects of both. It brings the basic Marxist concepts – developed to explain capitalist society – into the study of pre-capitalist societies …

Mode of production – foraging, feudal, capitalist
Means of production – hunting, fishing, horticulture
Relations of production – how these activities are organized

MARXIST ANTHROPOLOGY TAKES ECONOMICS AS FUNDAMENTAL TO HUMAN SOCIAL LIFE …

BUT IT ACCEPTS THAT MODES OF PRODUCTION IMPLY PARTICULAR SOCIAL RELATIONS - OFTEN POWER RELATIONS - AND ENTAIL CERTAIN SOCIAL FORMS AND CULTURAL CONSTRAINTS.

Marxism's Evolutionary View

Marxism is an evolutionary perspective that relies on the dynamic concept of **contradiction**. This means that a mode of production can break down and produce **transformation** to a more historically advanced mode.

Different modes of production are often "**in articulation**", meaning that pre-capitalist economies operate within and in relation to capitalist ones, and should be studied as such.

ANTHROPOLOGISTS INFLUENCED BY *MARXIST* IDEAS MADE THE SHIFT TO STUDYING "*OTHER*" CULTURES IN RELATION TO COLONIALISM AND GLOBALIZATION AND AS PART OF WORLD SYSTEMS.

YOU MEAN, THEY NOTICED HOW THE REAL WORLD WORKS?

Economic anthropology can now be summed up as the study of *The Social Life of Things*, the title of the 1986 collection of essays edited by Arjun Appadurai. It is less concerned with formal models, more with trying to describe and understand economic *actions* in their social and cultural context.

The Household Unit

The household is the primary unit of a society and the normal place to start fieldwork study.

Asking who is a member of the household, how it is organized, how it feeds itself and provides the goods and services it needs, brings together all the various fields of interest to the anthropologist from ecology, economy, family and kinship to wider social organizations such as politics, religion, customs and ritual symbolism.

The Forms of Family

There are different family forms in different societies.

Nuclear family – a married couple and their children.

Compound family – a man, his multiple wives or concubines, and all their children.

Joint family – a group of brothers and their wives and children living together.

Extended family – closely related nuclear families living together, including grandparents, parents and children. The term is also applied to such a group who may not live together but maintain close links, usually in an urban setting.

DIFFERENT FORMS OF FAMILY IMPLY DIFFERENT RIGHTS AND RESPONSIBILITIES THAT ARE RECOGNIZED BETWEEN MEMBERS OF THE FAMILY.

The Marriage Links

Marriage is the event that brings a family into existence. We can identify various types of marriage.

Monogamy –
one man and
one woman.

Polygamy –
one man with
more than
one wife.

Polyandry – one woman
with more than one husband,
usually a group of brothers.

ANTHROPOLOGISTS
HAVE ENCOUNTERED MANY
OTHER FORMS OF MARRIAGE ...

Ghost marriage –
marriage to a dead
person.

Levirate – marriage to
the sister of a dead wife.

Woman marriage –
marriage between
two women.

Marriage Contract Payments

In contracting a marriage, two kinds of payments may be made between the family or kin group of the bride and groom.

The household and domestic sphere were seen as the locale of women. **Phyllis Kaberry** (1910–77), who did fieldwork among the Australian Aborigines in the 1930s, depicted women as "active agents". Her work in Australia and West Africa has achieved new prominence with the rise of feminist anthropology and its focus on gender studies.

*W*OMEN ANTHROPOLOGISTS HAVE INCREASINGLY ARGUED THAT THERE HAS BEEN MALE BIAS AMONG ANTHROPOLOGISTS.

RE-EXAMINING THIS PROPOSITION HAS REVOLUTIONIZED THE WAY IN WHICH ANTHROPOLOGY CONCEIVES OF AND STUDIES THE HOUSEHOLD, GENDER, SEX, THE BODY, HUMAN RELATIONSHIPS AND THE SELF.

Phyllis Kaberry (1910–77)

The Study of Kinship

Kinship studies how people are related to each other, the different
ways in which kin relations are structured and the functions
performed by kin groups and kin status. Robin Fox declared in
Kinship and Marriage: *An Anthropological Perspective* (1967) …

(exists P ((P v Q) & (-P v R))) <-- v R)

(exists P ((P v Q) & (-P v R) v (-N S))) <-> ((Q v R) & (Q v S))

(exists P ((P v Q) & (-P v R))) (exists P ((P(a) v Q) & (-P(b) v R <-> (a = b -> (Q v R))

(exists P ((P v Q) & (-P v R) v (-P v S))) <-> ((Q v R) & (Q v S))

(exists P ((P(a) v Q) & (-P(b) v R))) <-> (a=b -> (Q v R))

KINSHIP IS TO
ANTHROPOLOGY WHAT LOGIC
IS TO PHILOSOPHY OR THE
NUDE TO ART; IT IS THE BASIC
DISCIPLINE OF THE SUBJECT.

IT'S
ALSO THE IDEA
THAT GAVE BIRTH TO
THE SUBJECT.

Kinship Codes

Kinship diagrams appear like a complex code, because that's exactly how anthropologists thought kinship operated – as the "code" that ordered and explained an entire society.

Kinship Symbols

F = Father	M = Mother	P = Parent
B = Brother	Z = Sister	G = Sibling
S = Son	D = Daughter	C = Child
H = Husband	W = Wife	E = Spouse

e = older (elder)   y = younger
ss = same sex   os = opposite sex

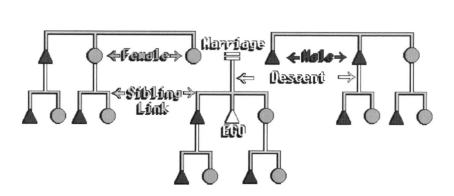

Kinship is more than, and distinct from, simple biology. Some societies do not consider biological conception as the only means to parenthood, for instance, as Malinowski argued …

Genitor – culturally recognized biological father
Pater – social father (including adoptive father)

Genetrix – culturally recognized biological mother
Mater – social mother (including adoptive mother)

Classificatory Kinship

DIFFERENT SOCIETIES RECOGNIZE AND NAME DIFFERENT KINDS OF RELATIVES.

THIS IS KNOWN AS CLASSIFICATORY KINSHIP TERMINOLOGY.

For example, all the children of a group of brothers may be classed as brothers and sisters. The classificatory relationship that a person occupies may be important in determining whom they can or should marry, and rights to inheritance, as well as various other family and social obligations.

Fictive Kinship

Apart from classificatory kin, there are also types of *fictive* kinship. *Godparents* are fictive kin, and may be important not only as ritual sponsors of children in Christian baptism but also as social sponsors of the child in later life, as well as additional sources of more relaxed relationships outside the tensions of the family.

Compadrazo is a fictive relationship between the child's godparents and parents, who may lend each other support or money, and co-operate in religious festivals or in times of trouble.

FICTIVE KIN RELATIONS MAY BE UNEQUAL.

PARENTS MAY SEEK FICTIVE KIN WHO ARE OF HIGHER STATUS AND THEREFORE ABLE TO ADVANCE THEIR OWN AND THEIR CHILD'S INTERESTS.

Mapping relationships builds to wider patterns and systems of kinship

Descent Theory

A kin group – or descent group – links people across generations according to lineal principles. This is the basis of *descent theory*. All of the people who trace lineal descent from a named common or *apical* ancestor will form a **lineage**. There are different principles for tracing descent and building lineages.

PATRILINEAL DESCENT IS DETERMINED THROUGH THE FATHER TO THE FATHER'S FATHER, FATHER'S FATHER'S FATHER AND SO ON ...

... DEPENDING ON HOW MANY GENERATIONS ARE RECKONED OR RECOGNIZED.

THE CHILD OF ANY MALE MEMBER OF THE GROUP, WHETHER MALE OR FEMALE, IS ALSO A MEMBER.

Matrilineal descent is traced through the mother to the mother's mother and mother's mother's mother. The descendants will be all those in the same female line. Matriliny is less common than patriliny, but exists in many societies in different parts of the world. Matriliny must be distinguished from **matriarchy**, the exercise of power and authority by women.

IN A MATRILINEAL SOCIETY, POWER AND AUTHORITY ARE STILL EXERCISED BY MEN.

A MOTHER'S BROTHER IS THE MALE HEAD OF THE MATRILINEAL DESCENT GROUP.

MATRILINY DOES NOT MEAN THAT THE FATHER IS NOT AN IMPORTANT FIGURE.

power

Double descent – in this system, a person belongs to two kin groups, one patrilateral and one matrilateral, a rather rare occurrence. It must be distinguished from *complimentary filiation*, obligations to kin on the opposite side of the kin system.

Cognatic or **bilateral** descent – a person is equally related to both the mother's and father's family, and there are no patrilineal or matrilineal groups.

Marriage and Residence Rules

THE KINSHIP STRUCTURE IS BASED ON MARRIAGE RULES – WHO IS A PROPER OR PREFERRED MARRIAGE PARTNER?

AND ALSO DICTATES RESIDENCE RULES – WHERE SHOULD THE MARRIED COUPLE LIVE? WITH HIS OR HER RELATIVES?

Virilocal residence – a wife will move to live in her husband's locality, which creates patrilateral kin groups.

Uxorilocal residence – the husband moves to the locality of the wife. This keeps matrilineally related women together.

Avunculocal residence – residence with the man's mother's brother creates villages of matrilineally related men.

The Idiom of Kinship

Kinship requires asking what rights and obligations, duties and responsibilities, go with specific named relationships. What kinds of behaviour are appropriate between people in such named and recognized kin relationships?

The idiom of kinship was seen by anthropologists to structure most political, economic and religious activities in society. It was the major institution of the society.

AS THE STRONGEST SOCIAL BOND, KINSHIP IS THE WAY PRIMITIVE SOCIETIES MAINTAIN ORDER AND CREATE SOCIAL SOLIDARITY.

OR YOU "FOUND" IT BECAUSE THAT'S WHAT YOU EXPECTED TO FIND, JUST AS THE HISTORY PREDICTED.

What's the "Use" of Kinship?

Unilineal descent theory and its consequences for the way in which society was studied and conceptualized became the subject of heated debate. In his 1966 article, "Ancestor Worship in Anthropology; Or Observations on Descent and Descent Groups", H.W. Scheffler argued that descent was used to highly varied ends among different peoples, indeed even by the same people, and not necessarily towards corporate group structure.

BUT ANOTHER CANDIDATE FOR UNDERSTANDING KINSHIP HAD ALREADY ENTERED ANTHROPOLOGY.

Alliance Theory and Incest Taboo

Alliance theory was originated by **Claude Lévi-Strauss** (1908–) in *The Elementary Structures of Kinship* (1949). It is the study of relations between groups, families or individuals through marriage. Alliance theorists reject the idea that descent groups are the basis of society, claiming instead that descent groups are "elements" in relations of marital exchange between groups.

THE INCEST TABOO IS THE BASIS OF CULTURE. ANTHROPOLOGISTS HAVE FOUND IT TO BE UNIVERSAL, AND YET IT OPERATES DIFFERENTLY IN DIFFERENT CULTURES.

THE INCEST TABOO IS THE PROHIBITION ON SEXUAL RELATIONS BETWEEN PEOPLE IN SPECIFIED KIN RELATIONSHIPS.

In societies in which all people are "kin", there may be categories of kin with whom it is permissible to marry and have sexual relations.

Structures in Mind

Lévi-Strauss was interested in the patterns or "structures" that people have in mind when they operate social rules – in this case, marriage rules. He argued that elementary structures represent the earliest forms of human kinship.

Elementary structures have positive rules of marriage – or the opposite of incest taboos.

YOU MUST MARRY A "CROSS-COUSIN", FOR EXAMPLE.

Complex structures have negative rules.

YOU CANNOT MARRY YOUR SISTER.

The **Crow–Omaha system** stands between elementary and complex and includes most, though not all, societies with Crow or Omaha kinship terminologies.

WHILE THIS SYSTEM DEFINES WHOM ONE CANNOT MARRY (COMPLEX), THERE ARE SO MANY PROHIBITIONS THAT IN PRACTICE IT RESEMBLES ELEMENTARY STRUCTURES.

Forms of Elementary Structures

Generalized exchange – Group A give wives to Group B, who give wives to Group C.

Delayed restricted/Delayed direct exchange – women move in one direction between groups and in the next generation in the opposite direction.Very rare. Anthropologists argue that it exists only as a theoretical possibility.

Restricted/Direct exchange – Group A give women to Group B and Group B give women to Group A.

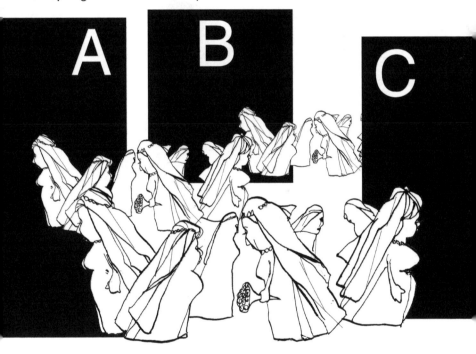

Types of marriage rules are:
Matrilateral cross-cousin marriage: a man marries his mother's brother's daughter.
Patrilateral cross-cousin marriage: a man marries his father's sister's daughter.
Moieties: literally meaning "halves", it refers to a society with just two descent groups that exchange marriage partners.

Does Alliance Theory Work?

Alliance theory led to heated debates. Increasingly, it has been shown that marriage rules are flexible and can be used to highly varied ends among different peoples, indeed even by the same people.

FEW IF ANY FEATURES OF SOCIAL ORGANIZATION CAN BE PREDICTED FROM KNOWLEDGE OF THE MARRIAGE RULES.

RULES

YOU SEEM TO RUN INTO THE PROBLEM OF REALITY QUITE A LOT. PERHAPS THAT SHOULD TELL US SOMETHING!

In *A Critique of the Study of Kinship* (1984), David Schneider advises anthropologists to stop looking for "kinship" because it is a vacuous and confused domain.

So what happened to kinship studies? It remains an important topic, but kinship is no longer *the* study that explains all. Kinship is of interest to those who study the social organization of human reproduction. It is of interest to those who study how sexual relations, definitions of personal identity and gender roles are culturally constructed.

A whole new array of language has entered the topic. Such terms as *self*, *agency*, *gender*, the *life of values* and *affect*, *personhood*, have become part of anthropology. This new terminology signals a shift in focus.

Politics and Law

The classic view of kinship stressed its role in the maintenance of social order. It naturally led to the study of politics and law, the structure of authority, power, control and decision-making in a society. There are two basic approaches to studying politics. The first is the **typological approach** which classifies types of polities and correlates political organization with the *subsistence base* and *patterns of kinship*. Here are basic examples …

Band societies are usually hunter-gatherers (but may also be fishermen or horticulturalists) whose social structure is based on kinship.

WE HAVE AN EGALITARIAN WAY OF LIFE AND NO EMPHASIS ON LEADERSHIP.

LEADERSHIP MAY BE TEMPORARY AND FOR SPECIFIC PURPOSES – A HUNTING PARTY OR WAR PARTY.

Tribal societies are usually pastoralist (livestock-based) or horticulturalist. The social structure is based on clans and lineages; age and gender may be important factors, as in "age grade" societies (see page 102). They may be acephalous (literally "headless"), meaning without leaders.

Further Examples ...

Chiefly societies have economies based on livestock, horticulture or intensive agriculture. They are governed by hereditary chiefs who have power, authority and often inherited wealth.

State societies have economies based on intensive agriculture and often a developed market system with extensive trade networks, both internal and external. With a high population density, they may be divided by class or other means of social stratification, for example on caste principles.

THEY MAY HAVE HEREDITARY OR ELECTED LEADERS WHO WIELD POWER AND AUTHORITY.

SUCH LEADERS HAVE SACRED DUTIES OR SUPERNATURAL POWERS, AS IN SOME AFRICAN KINGDOMS.

The Terminological Approach

The typological approach clearly includes the idea of an evolutionary development of political structures from simple to complex. It is particularly associated with **Elman Service** (1915–96) and his book, *Profiles in Ethnology* (1978).

Such terminology can be applied to the analysis of all political systems. **M.G. Smith** characterizes this approach in his book *Government in Zazzau*, published in 1960.

Political Anthropology

Political anthropology examines and compares diverse systems of social control, power structures, the extent of consensus and patterns of equality and inequality, how leaders establish or bolster their authority through tradition, force, persuasion and religion.

Age Grade Societies

Studying the tension between generations and the mechanisms for securing the succession from one age grade to another played an important part in the development of the **processual approach** of **Max Gluckman** (1911–75). His classic studies are *Custom and Conflict in Africa* (1955) and *Politics, Law and Ritual in Tribal Societies* (1965). This approach distinguished the members of the **Manchester School** headed by Gluckman and based at Manchester University.

Synchronic vs. Diachronic Views

Early functionalism and structural functionalism developed by Malinowski and **A.R. Radcliffe-Brown** (1881–1955) saw politics as embedded in kinship. These approaches produced a static view of society that emphasized **synchronic** practice.

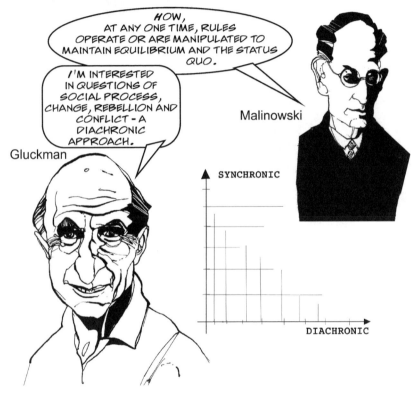

HOW, AT ANY ONE TIME, RULES OPERATE OR ARE MANIPULATED TO MAINTAIN EQUILIBRIUM AND THE STATUS QUO.

I'M INTERESTED IN QUESTIONS OF SOCIAL PROCESS, CHANGE, REBELLION AND CONFLICT – A DIACHRONIC APPROACH.

Malinowski

Gluckman

SYNCHRONIC

DIACHRONIC

Gluckman developed a **diachronic** understanding of the way in which social processes operate through time. He distinguished between …

rebellion – displacing people in power
revolution – displacing or changing the system in which power operates

Gluckman suggested that rebellion was a permanent process in political systems in which instability is normal.

Other Social Stratifications

CLASS SOCIETIES ARE STRATIFIED BY UNEQUAL ACCESS TO POLITICAL POWER AND ECONOMIC GOODS.

CASTE SOCIETIES ARE STRATIFIED BY RITUAL, RELIGIONS AS WELL AS BY SOCIAL AND ECONOMIC INEQUALITIES.

The caste system of India was the subject of the classic study *Homo Hierarchicus* (1967) by **Louis Dumont** (1911–98).

Transacting Identity

The tension between these strategies led anthropologists to an interest in **ethnicity** and **nationalism**.

Fredrik Barth (1928–), especially in his edited volume *Ethnic Groups and Boundaries* (1969), introduced the concept of **transactionalism**. Barth's earlier study, *Political Leadership Among Swat Pathans* (1959), showed how leaders maintain allegiance by transaction and a constant "game" oscillating between conflict and coalition. The concept was developed to study the negotiation of identity in the study of ethnicity.

Problems of Ethnicity

Ethnicity is generally concerned with the way in which groups set themselves apart and derive a sense of "we" or "us". Anthropologists are concerned with the various ways in which people express these differences, and how they are experienced. It is distinguished from **race**.

RACE SETS OTHER GROUPS APART AND DISTINGUISHES BETWEEN "US" AND "THEM" BY STEREOTYPING OTHERS AND THEIR PRACTICES.

THE LIKELY CONSEQUENCES ARE RACISM, DISCRIMINATION AND VIOLENCE.

THEM

US

Alterity – the concept of *alien-objectified otherness* – is a recent arrival in anthropology. It suggests that all societies and groups have notions of alterity. But it also includes a "self-reflexive" debate about the history and practice of anthropology itself.

Colonialism

The study of **colonialism** – the exercise of political domination and control by one society over another – is another late developer in anthropology. The Manchester School and Rhodes-Livingstone Institute have produced studies of social change and the distinction between tribal and town life that make the existence of colonialism visible and pertinent to anthropological concerns, rather than invisible and irrelevant.

Other studies have looked at the ways in which Western legal systems have been adopted and adapted by non-Western people.

Anti-capitalist Anthropology

Marxist-inspired anthropologists concerned about colonial and post-colonial struggles against capitalism and the state have produced new perspectives on the intertwining of culture and politics. New concepts and terminology entered the field.

Centre–Periphery is a concept developed by **Immanuel Wallerstein** (1930–) in *The Modern World System* (3 vols., 1974–89).

Dependency was developed by **André Gunder Frank** (1929–) in *Capitalism and Underdevelopment in Latin America* (1967).

THE CENTRE IS THE PLACE WHERE POWER IS EXERCISED. THE PERIPHERY, THE LOCATION AFFECTED BY DECISIONS MADE AT THE CENTRE.

THE DEVELOPMENT OF CAPITALISM WAS DEPENDENT UPON COLONIAL EXPLOITATION AND CREATED DEPENDENCY, IMPOVERISHMENT AND IMPEDIMENTS TO DEVELOPMENT IN THE COLONY.

Gunder Frank

Anthropologists have developed the idea of the **global** and **local** in cultural terms. Studying the everyday practices of subordinated groups, anthropologists looked at informal structures: **interstitial**, **supplementary** and **parallel** forms of action demonstrated by **coalitions**, **cliques** and **networks**.

The concept of the **invention of tradition**, derived from a collection of essays of the same name edited by Eric Hobsbawm and Terence Ranger (1966), has been used by anthropologists to explore the **politics of culture** in various nationalisms.

IN WEAPONS OF THE WEAK (1985), JIM SCOTT LOOKED AT POLITICS AS MODES OF EVERYDAY RESISTANCE ...

THIS WAS THE BEGINNING OF A TREND TO STUDY MINORITY MOVEMENTS WITH A FOCUS ON THE POLITICS OF VIOLENCE AND RESISTANCE TO STATE RULE.

Anthropology of Law

Anthropologists distinguish between **law** and **custom**, but have shown that in operation there is little difference between the two concepts.

Paul Bohannan in *Law and Warfare* (1967) distinguished law from custom and rules of conduct, as being "doubly institutionalized" – meaning that law re-institutionalizes customs or rules derived from other institutions. Law is "a custom that has been restated in order to make it amenable to the activities of the legal institutions".

American anthropologist E. Adamson Hoebel suggested in *The Law of Primitive Man* (1954) that law entails three principles.

1. The legitimate use of force to ensure correct behaviour and punish wrongdoing.

2. The allocation of power to individuals to use coercion.

3. Respect for tradition as against whim. Enforcement must be based on the existence of known rules, whether customs or statutes.

Mechanisms for Resolving Disputes

Avoidance occurs in hunter-gatherer societies.

SOCIAL SPACE IS GREAT AND FORMAL MECHANISMS OF CONTROL ARE RELATIVELY UNDEVELOPED.

DIVINATION OR ORDEALS CAN BE USED TO DISCOVER THE SOURCES OF CONFLICT AND AGGRESSION BETWEEN PEOPLE.

Mediation, **Negotiation**, **Arbitration**, **Adjudication** lead to a variety of arrangements that go beyond settling **conflict** or **contention** – tension that is endemic in the social fabric. They deal with resolving **disputes**, a specific incident of contention, and may involve settling compensation or operating customary sanctions.

There may be specific, often ritualized, groups invested with authority to arbitrate and adjudicate disputes. Or disputes may be dealt with by formally constituted courts.

Religion

The study of religious beliefs is a major area of interest for anthropologists, who have categorized many different forms of religious organization and practice.

ANIMISM IS THE BELIEF IN SPIRITS INHABITING NATURAL PHENOMENA SUCH AS MOUNTAINS, RIVERS OR TREES.

E.B. TYLOR, LIKE MOST EARLY ANTHROPOLOGISTS, ARGUED THAT THIS WAS THE OLDEST FORM OF RELIGION.

FETISHISM WAS ANOTHER INTEREST OF EARLY ANTHROPOLOGISTS, WHO THOUGHT THAT "PRIMITIVE" PEOPLE MADE OBJECTS BELIEVED TO HAVE MAGICAL POWERS.

Although fetishes exist, they do not form the basis of a belief system.

Totemism is an Ojibway word from the Great Lakes region of North America. Totems are spiritual entities represented by animal species. A particular totem symbolizes a clan. People who share the same totem cannot marry. An individual may have a totem – a personal guardian spirit that can be associated with a food prohibition: "you must not eat the animal that represents your totem". Another form of totem belongs to the spirits of sacred sites.

In his book *Totemism* (1962), Lévi-Strauss argued that there is no such thing as totemism.

THE TERM APPLIES NOT TO A SINGLE PHENOMENON, BUT SEVERAL.

THE MAJOR DISTINCTION IS BETWEEN "TOTEMS" THAT ARE EMBLEMS OF CLANS OR OTHER SOCIAL GROUPS, AND "TOTEMS" THAT ENTAIL FOOD PROHIBITIONS AND A RANGE OF SACRED ASSOCIATIONS.

Shamanism and Cargo Cults

A shaman is a religious adept who mediates between the human world and the spirit world, between humans and animals, or between the living and the dead.

IT IS POLITICALLY CORRECT TERMINOLOGY FOR "MEDICINE MAN", "WIZARD" OR "WITCH-DOCTOR".

CARGO CULTS AND MILLENARIAN MOVEMENTS ENVISION THE END OF THE WORLD OR THE DAWN OF A NEW AGE.

Cargo cults are named after movements studied in Melanesia which were especially prevalent after the Second World War. The cargo refers to valuable Western goods. In some cases, prophets declared the imminent return of ancestors with "cargo" that in the new age would be controlled by the natives themselves and not the white man. Similar movements in North America are termed **nativistic** or **revitalization** movements.

Two other basic forms of religion are: **polytheism** – belief in the existence of more than one deity; and **monotheism** – belief in the existence of only one God. *The Elementary Forms of Religious Life* (1912), by the French sociologist **Emile Durkheim** (1858–1917), is the source of two perspectives on religion. The first is a **functionalist** perspective: religion, as belief and action, is defined by what it does.

RELIGION IS A *SOCIAL* CREATION THAT REINFORCES SOCIAL SOLIDARITY.

THE SECOND IS THE BASIC DISTINCTION BETWEEN THE "SACRED" AND "PROFANE", AS WE'LL SEE NEXT ...

Emile Durkheim

Sacred and Profane

The **Sacred**, set apart from the normal world, includes hidden, forbidden or special knowledge and practices, such as taboos associated with ritual.

IT IS ASSOCIATED WITH MAGICAL FORCES, SPIRITS OR DEITIES AND CAN BE RELATED TO BOTH RELIGION AND MAGICAL PRACTICES.

The **Profane**, belonging to the everyday world, entails everyday knowledge and includes utilitarian practices.

IT IS ASSOCIATED WITH ORDINARY LIFE, ESPECIALLY MATERIAL THINGS RELATED TO SECULAR ACTIVITIES.

The Anthropology of Magic

Belief systems may also include witchcraft and sorcery, sometimes jointly termed as **magic**.

WITCHCRAFT INVOLVES A MALEVOLENT MAGICAL POWER INHERENT IN THE MAKEUP OF AN INDIVIDUAL.

SORCERY INVOLVES A MALEVOLENT POWER THAT IS LEARNED RATHER THAN INHERITED.

E.E. Evans-Pritchard (1903–73) produced the seminal study: *Witchcraft, Oracles and Magic Among the Azande* (1937). The Azande are obsessed with witchcraft. Evans-Pritchard shows that this belief has its own logic and reasonableness. It operates as another layer beyond rationalist cause and effect to explain personal fortune and misfortune.

The Debate on Belief

Evans-Pritchard's study became important in the "what is it reasonable to believe?" debate. The debate questioned the meaning of **reason** and **rationality**. It explored the concept of **cultural relativity** – understanding ideas, beliefs and practices in their own cultural context from the perspective of the believer and practitioner.

*IT HIGHLIGHTED **ETHNOCENTRISM** - THE PRIVILEGING OF **WESTERN** VIEWPOINTS - ESPECIALLY SCIENTIFIC, INSTRUMENTAL RATIONALITY AS THE SOLE DETERMINANT OF "REALITY".*

Evans-Pritchard

Examining Ritual

Religion involves **ritual** and may be expressed through **myth** and **symbolism**, all of which may inform and be part of **art**. Rituals may mark major social events or stages in individual life cycles. In rituals, belief systems are performed, symbolically enacted and reinforced, and so operate to assert or construct meaning both individually and collectively.

RITUALS EXPRESS IDENTITY, INDIVIDUALLY OR COLLECTIVELY ...

THEY CAN SERVE TO RESOLVE OR RELEASE SOCIAL TENSIONS AND CONFLICT AND PROMOTE SOCIAL COHESION OR SOLIDARITY.

Rites of Passage

Arnold Van Gennep (1873–1957) wrote the classic study of rites of passage – rituals that mark major events in an individual's life cycle. These occasions can be named as follows …

Naming rites mark the transition from non-person to person – from person outside the community to member of the community.

Initiation rites mark the transition from one status to another, especially from childhood to adulthood.

Marriage rites mark the transition from single to married status.

Funeral rites mark the transition from person to ancestor – from the present world to the beyond.

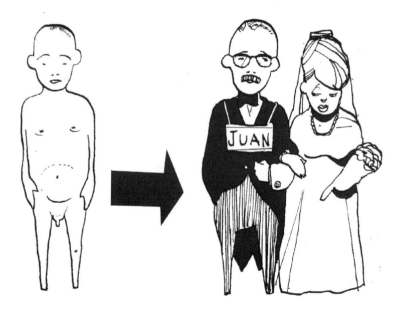

All rites of passage involve three stages. **Separation**: leaving the group prior to rituals, creating spatial and symbolic distance between the status before and the status acquired during the rite. **Transition** – the *liminal* phase (from the Latin for "threshold") – the period of most ritual activity. The period of transition and becoming can be either dangerous or formative.

IT MAY INVOLVE REVERSAL OF NORMALITY – THE OVERTURNING OF ACCEPTED BEHAVIOUR OR CATEGORIES.

IT MAY CREATE SPECIAL BONDS BETWEEN INDIVIDUAL MEMBERS OF A GROUP WHO JOINTLY UNDERGO THE RITUAL.

IT MAY INVOLVE THE TRANSFER AND ACQUISITION OF NEW KNOWLEDGE.

Incorporation, at the completion of the rite, is the reintroduction to society of individuals who have attained a new status.

The Study of Myth

Myths are tales – sacred or religious in nature – social rather than individual or anecdotal in subject matter, and concerned with the origin or creation of phenomena whether natural, supernatural or socio-cultural. Myths may be acted out in particular rituals. Myths and rituals share common symbolic elements and are complementary aspects of creative and religious expression.

BOAS SAW MYTH AS A REPOSITORY OF INFORMATION ABOUT CULTURE AND CULTURE TRAITS, AS WELL AS A GUIDE TO REGIONAL RELATIONS BETWEEN GROUPS.

MALINOWSKI SAW THEM AS A "CHARTER FOR SOCIAL ACTION" - RATIONALIZATIONS USED TO EXPLAIN AND JUSTIFY WHAT PEOPLE DO, THEIR CUSTOMS AND BEHAVIOUR.

Claude Lévi-Strauss

The French anthropologist most associated with the study of myth, **Claude Lévi-Strauss** (1908–), initially studied law and philosophy. In 1934, he went to Brazil to teach sociology but ended up doing fieldwork among the Bororo Indians.

MYTH IS A TYPE OF THOUGHT – AN EXAMPLE OF UNIVERSAL "STRUCTURAL PRINCIPLES" THAT UNDERLIE ALL HUMAN CULTURAL AND SOCIAL SYSTEMS.

Myth is used to reflect on and symbolically mediate or resolve universal and culturally specific contradictions or *oppositions*. Oppositions are especially important in Lévi-Strauss's structuralist system.

Binary Oppositions and Structure

Oppositions are **binary**: death/creation, maternal/paternal, raw/cooked. Myths endlessly combine and recombine the different symbolic elements. The different versions of myths demonstrate constant creation and modification of mythical knowledge and thought. For Lévi-Strauss, the essence of culture is structure, each culture having its own configurations or structures. These structures exist as part of a worldwide system of all possible structures founded on the psychic unity of humankind.

MY CONCERN IS WITH THE IDEAL STRUCTURES OF SOCIETY. THE ANTHROPOLOGIST WORKS OUT ABSTRACTLY ALL THE POSSIBLE PERMUTATIONS.

SO, STRUCTURES COME IN TWO FORMS: WHAT IS IN THE ANTHROPOLOGIST'S MIND AND WHAT IS IN THE MIND OF THE PEOPLE STUDIED. WHOSE IDEAS MATTER MOST?

Symbols and Communication

Symbols are central to the discussion of ritual and myth and raise questions of **meaning** and **communication**. The study of symbolism has developed different approaches such as **symbolic anthropology** and **cognitive anthropology.** Concepts and terminology are borrowed from **linguistics** and **semiotics.**

BUT MUCH CAUTION IS NEEDED, BECAUSE THEY CAN BE USED IN OPPOSITE WAYS BY DIFFERENT WRITERS.

YOU MEAN, THEY MAKE IT UP AS THEY GO ALONG?

Symbols and the Social Process

Victor Turner (1920–83), a leading member of the Manchester School, focused on symbols as part of social process. He argued, "we master the world through signs … we master … ourselves through symbols."

Turner introduced the concept of **communitas**, the primal ground or creative impulse of a culture that is accessed through symbolism.

Actor, Message and Code

Anthropological approaches to symbolism emphasize the **actor** – the person employing or engaging with symbolism – rather than the **message**, and the message rather than the **code**.

Gregory Bateson (1904–80) saw cultures as a mechanism for the generation and transmission of information. He introduced the concepts of **play** and **metacommunication**. Play and creativity in symbolism and ritual are activities through which people expand and reorganize their consciousness.

Symbolism and New Perspectives

The study of symbolism has been important in recent developments in anthropology.

The **symbolic anthropology** of **David Schneider** (1918–95) takes culture as a total system of meanings and symbols. Symbolic systems should not be separated into bits and linked to particular aspects of social organization – economy, politics, kinship, religion – but studied as wholes.

Cognitive anthropology borrows from linguistics the distinction between sounds – *phonetics*, and meaningful units of sound – the *phonemic*. This became the **etic** – units of any kind – and the **emic** – meaningful units of any kind. The etic is the level of universals apparent to an "objective" observer, while the emic is the meaningful contrasts within a particular language or culture. In this way, culture is seen as an ideational system, a system of knowledge and concepts.

Interpretive anthropology, which began with Evans-Pritchard's work on Azande witchcraft and Nuer religion, has come to be most associated with the work of the American anthropologist **Clifford Geertz**, (1926–). Geertz proposed the study of cultural systems as **texts**, or **acted documents**, to be studied by building up the details of cultural life as **thick description**, a methodology of doing ethnography.

Geertz criticized what he called Lévi-Strauss's "cerebral savages" and his "cryptological" approach that analysed symbols as closed structures rather than texts built out of social materials.

MEANING DERIVES FROM PURPOSE AND NOT FORMAL STRUCTURES. LÉVI-STRAUSS'S FOCUS ON INTERNAL RELATIONS OF SYMBOLIC ELEMENTS DISTRACTS FROM THE INFORMAL LOGIC OF ACTUAL LIFE.

In his 1966 article, "Religion as a Cultural System", Geertz defined religion as: "a system of symbols which acts to establish powerful, pervasive and long-lasting moods and motivations in men by formulating conceptions of a general order of existence and clothing these conceptions with such an aura of factuality that the moods and motivations seem uniquely realistic."

Anthropology of Art

Religion, belief, ritual and symbolism are linked to another major interest of anthropologists: **art**. The anthropology of art has been most concerned with material objects such as sculpture, masks, paintings, textiles, baskets, pots, weapons and the human body itself. These are seen not merely as **aesthetic** objects, appreciated for their beauty, but as playing a wider role in people's lives.

In many societies, the artist is not specially recognized as an "individual creator", nor are their works distinguished as part of a separate **high culture**. Artistic production is often general to large numbers of people, being collective rather than individual.

Visual Anthropology

Another recent development is visual anthropology. The study of visual systems has expanded to include the study of local photographic practice and local television and film production.

It is most accessible as the **ethnographic film**, a visual record of events, rituals, activity and setting studied and written about by anthropologists. This includes visual materials recorded and produced by anthropologists in the course of doing fieldwork.

BUT IT ALSO INCLUDES THE COLLABORATION OF ANTHROPOLOGISTS WITH PROFESSIONAL TELEVISION COMPANIES AND FILM-MAKERS.

AND IT MAKES ANTHROPOLOGISTS FAMOUS!

Disappearing World

The television series *Disappearing World* (1970–) is a classic example of ethnographic film. Some films have become notorious, such as *The Feast* – Napoleon Chagnon and Timothy Asch's film of the Yanomamo Indians of the Venezuelan rainforest.

A film raises lots of questions. Were the events staged? What influence and effect did the presence of a film crew have on the event being recorded? Does film privilege the emotive impact of visual imagery on the audience over analysis?

A New Branch or an Old Root?

Some anthropologists argue **applied anthropology** and the related field of **development anthropology** should be recognized as new branches of the discipline. The debate is whether "development" is nothing more than a recurrence of the colonial relationship – "underdeveloped/developed" being simply the reformulation of "primitive/civilized".

ANTHROPOLOGY HAS BECOME PART OF THE PRACTICE AND THEORY OF DEVELOPMENT AGENCIES ...

... OF VOLUNTARY ORGANIZATIONS, INTERNATIONAL ORGANIZATIONS AND GOVERNMENTS.

It has become part of the practice of governmentality – "anthropology has a long history of wanting to make itself useful to governments", writes Henrietta Moore.

Writing Up the Field

Once the data has been gathered in the field, it must be written up. The standard means of presenting one's fieldwork is the **ethnographic monograph**. Classic ethnographies come in a number of forms.

Seamless narrative – detailed, often very long and weaving all the details of social life together without organizing them into discrete topic areas.

Life cycle ethnographies – structured on the progression from childhood to old age, using each stage in the life cycle to organize the presentation of aspects of social life, ritual and belief.

Structured by social systems – material organized according to the heading of environmental setting, economics, politics, law and social control, kinship, ritual and belief. Typically concluding with a chapter on social change.

Writing in the Present

The principal device in classic ethnographic monographs is to write in the **ethnographic present**. This is more than merely writing in the present tense. It can be to write **ahistorically**, that is, to present a perspective on culture and the lifeways of a people as if time and change, or outside influence, do not exist.

The ethnographic present is a frozen portrait conveying the conclusion that everything within a culture fits and functions to keep it *in stasis* – a stable ongoing equilibrium that endlessly reproduces the same pattern.

An ethnographic monograph may simultaneously be presenting a portrait of one particular culture, while advancing arguments about theoretical problems in anthropology. The fieldwork site is chosen because it is considered likely to produce material pertinent to a particular theoretical interest, to provide evidence to answer a particular theoretical question.

OR BECAUSE IT'S A GOOD PLACE TO INVENT AND PROVE YOUR PET THEORY!

Auto-Anthropology

Older ethnographies are often easy to read, fascinating and entertaining. The more modern the ethnography, the more turgid, jargon-laden, "terminologically challenged", self-absorbed and impenetrable it tends to be.

WHEN YOU REACH **AUTO-ANTHROPOLOGY** - THE ETHNOGRAPHY THAT IS ALMOST AN AUTOBIOGRAPHY OF THE ANTHROPOLOGIST - YOU KNOW YOU ARE IN THE PRESENT DAY.

For its first 50 years, modern anthropology was happy doing ethnography and whittling away the old roots. Then a series of debates began that altered the discipline. The first of these was the debate over the Mexican village, Tepoztlan.

The Dual/Duel of Tepoztlan

In 1930, **Robert Redfield** (1897–1958) published *Tepoztlan: A Mexican Village*. Redfield combined Boasian functionalism with evolutionist and German sociological traditions to focus on the normative rules that governed social behaviour. He produced an idealist representation of a village where people lived in peaceful harmony.

REDFIELD BECAME A THEORIST OF PEASANT SOCIETY.

I DEVELOPED THE CONCEPT OF THE GREAT AND LITTLE TRADITIONS - THE URBAN-FOLK CONTINUUM.

Robert Redfield

Redfield's distinction is between the literate, urban (*Great*) culture of the élite and the largely oral and informal (*Little*) tradition of the peasant community. Elements from the Little tradition are constantly being taken up and reworked by the Great tradition. In time, these filter back down to be reinterpreted or transformed in the folk tradition, in accordance with local customs and values.

Tepoztlan Revisited

Tepoztlan was revisited by **Oscar Lewis** (1914–70). In *Life in a Mexican Village: Tepoztlan Restudied* (1951), Lewis used a processual approach that focused on behaviour itself – which turned out not to conform to Redfield's rules.

I FOUND A VILLAGE FULL OF FACTIONALISM, PERSONAL ANTAGONISM, DRUNKENNESS AND FIGHTING.

LEWIS WENT ON TO DEVELOP THE CONCEPT OF THE CULTURE OF POVERTY.

HIS BOOKS ABOUT TEPOZTLAN ARE CLASSICS, BOTH HIGHLY READABLE AND POPULAR.

Oscar Lewis

Question: What accounts for this difference – the almost diametrical inversion? The irreducible difference is the two anthropologists. Their views of the village were not merely *theory-driven* but related to radically different orientations.

Is Anthropology a Science?

In a series of radio lectures, published as *Social Anthropology* (1951), Evans-Pritchard questioned the assumption that anthropology is a science.

Long before postmodernism, Evans-Pritchard developed the idea of **translation of culture**. He meant getting as close as possible to the collective mind and thought of the people studied and then translating the alien ideas of one culture into equivalent ideas within Western culture. Which is what historians do when they study the past.

If anthropology is not a science but a branch of the humanities, what of its authority?

A Pretended Science

Science is supposedly objective, value-neutral empirical inquiry, and as such the arbiter of authority in Western society. Anthropology's claims to inclusion within this privileged domain were questioned but not radically shaken by Evans-Pritchard.

"We can pretend that we are natural scientists collecting unambiguous data and that the people we are studying are living amid various unconscious systems of determining forces of which they have no clue and to which only we have the key. But it is only pretence."

Paul Rainbow (*Reflections on Fieldwork in Mexico*, 1977)

The Indians are Off the Reservation

Vine Deloria Jr., a Lakota (Sioux) lawyer, became executive director of the National Congress of American Indians in 1964. Five years later, Deloria published *Custer Died for Your Sins: An Indian Manifesto* which posed some fundamental questions. "Why should we continue to be the private zoos for anthropologists? Why should tribes have to compete with scholars for funds when the scholarly productions are so useless and irrelevant to real life?"

Perhaps we should suspect the real motives of the academic community. They have the Indian field well defined and under control. Their concern is not the ultimate policy that will affect the Indian People but merely the creation of new slogans and doctrines by which they can climb the university totem pole.

At the annual meeting of the American Anthropological Association (AAA) in 1972, Deloria presented his case to the members.

Who Speaks for the Indian?

Twenty years later, the AAA held a retrospective, published as *Indians and Anthropologists* (1997). The conclusion of the book is an essay by Deloria assessing the impact of his diatribe against anthropology over a period of 28 years. "Anthropology continues to be a deeply colonial discipline. We still find it more valuable to have an Anglo know these things and be certified to teach them to other Anglos in an almost infinite chain of generations than to change the configuration of the academic enterprise and move on to more significant endeavours."

White Man as God

Similar questions of authority arise in the "Captain Cook" dispute between the leading American anthropologist **Marshall Sahlins** (1930–) and Singhalese anthropologist **Gananath Obeyesekere** (1930–). It has generated a small library of its own.

THE ISSUE IS WHETHER NATIVE HAWAIIANS REALLY MISTOOK ME FOR THEIR RETURNING GOD, LONO, IN 1779, AND RITUALLY AND APPROPRIATELY SACRIFICED ME.

Captain Cook

OR WHETHER THE "WHITE MAN AS GOD" MYTH IS A CONSTRUCT OF *WESTERN* SOCIETY.

This myth has been visited repeatedly on Hawaiians and substantiated by anthropology through its concepts of primitive thought and mentality as non-rational, pre-logical and superstitious, as Obeyesekere argues.

The Myth of Authority

The white man as god is foundational to Western thought. The "white god" is Cortes in Mexico and Pizarro in Peru. It was implicit in the 1607 instructions of the Virginia Company – to prevent the natives being aware of any injury or death to a white man – mortality and godhood being incompatible. Sahlins fulminates that Obeyesekere makes Hawaiians nothing other than "Enlightenment rationalists". His seeming mastery of the Hawaiian sources argues for the authority of anthropology as objective and privileged knowing.

Event Horizon

While Vine Deloria was blasting away at anthropology from one side, a group of radical American anthropologists were launching their own assault. *Reinventing Anthropology* (1969), a collection of essays edited by Dell Hymes, was motivated by the general political climate of the 1960s – the war in Vietnam, civil rights issues and protest at home in the United States.

IT CALLED FOR A PROGRAMME OF REFORM IN ANTHROPOLOGY. BUT THAT DIDN'T CATCH ON.

ITS CRITIQUE OF ANTHROPOLOGY HAS BECOME PART OF THE CONTEMPORARY DISCOURSE OF ANTHROPOLOGY.

Self-critical Anthropology

Reinventing Anthropology's reform elements have led to multiple interpretations and approaches.

Reporting back – meant not only the study of the impact of colonialism on the Other, but studying its consequences within the West, a rather different approach to that of routine anthropology within Western society.

Reflexive anthropology – making the process of studying visible by a change in methods of recording and writing field data to enable the voice of the studied to be heard and let them speak for themselves. It has become the self-scrutiny of the anthropologist.

Advocacy anthropology – the switch from disengaged to engaged study, participation with the economic, political, human rights and land rights predicaments of the peoples studied by anthropologists. But this leads to the survival debate …

A Hero of Anthropology

The American **Margaret Mead** (1901–78) is one of the most famous and widely read anthropologists. Her books *Coming of Age in Samoa* (1928) and *Growing up in New Guinea* (1930) remain standard introductory texts in anthropology and other social-science disciplines. Mead herself was an influential popularizer, commentator and established luminary on both sides of the Atlantic.

I WAS A STUDENT OF *BOAS* AND MARRIED TO *GREGORY BATESON* - ANOTHER INFLUENTIAL ANTHROPOLOGIST.

THE EXPERIENCES OF THEIR CHILDREN PROVIDED THE RESEARCH BASE.

Margaret Mead

Gregory Bateson

Their ideas had a crucial influence on Dr Benjamin Spock's theories of child-rearing that became the essential handbook of 1960s and 70s parents.

The Fall of the Mead Myth

In 1983, Derek Freeman published *Margaret Mead and Samoa: The Making and Unmaking of an Anthropological Myth*, in which he made a number of charges. Mead's investigation of Samoa was theory driven. She set out to prove the theory of her teacher Boas on the primacy of *nurture* (culture) over *nature* (biology). Her research in Samoa violated the ethnographic practice of her teacher. It consisted of sitting on a missionary veranda and being visited by four adolescent Samoan girls.

IN THESE INTERVIEWS, WE SHARED OUR SEXUAL FANTASIES WITH *MEAD*.

I BELIEVED THEY WERE "REAL" AND MADE THEM THE BASIS OF MY ANALYSIS OF *SAMOAN SOCIETY*.

A great hero of anthropology was mercilessly exposed. Defenders of Mead agree that Freeman establishes her representation of Samoa as false.

153

Observers Observed

The fall of Margaret Mead stands on another event horizon in anthropology. In the same year (1983), *Observers Observed: Essays on Ethnographic Fieldwork*, edited by George Stocking Jr., was also published. It was the first volume in the continuing series *History of Anthropology*. *Observers Observed* examined ethnographic practice as constructive acts by anthropologists. Stocking's own essay locates the shock of discovering Malinowski's field diary, first published in 1967, in a more general critique. In his diary, Malinowski recorded his longing for white civilization while in the field.

Feet of Clay

The history of anthropology properly concerns, Stocking suggests, the "background of historical experience and cultural assumptions that provoked and constrained it, and by which it was conditioned".

"The anthropologist as hero" was a phrase coined by the American critic and writer Susan Sontag (1966). It should be investigated so that the "hero's" feet of clay become evident.

THE LINK BETWEEN ANTHROPOLOGY AND THE BROADER EUROPEAN TRADITION OF "PARTICIPATORY EXOTICISM" CALLS FOR EXAMINATION.

Susan Sontag

The Issue of Self-projection

The "intrepid anthropologist" myth received another hammer blow from **Sir Edmund Leach** (1910–89), one of the most senior and leading exponents of British anthropology. In *Glimpses of the Unmentionable in the History of British Social Anthropology* (1984), Leach accepts that every anthropologist can be expected to recognize in the field something no other observer will see, a projection of his or her personality.

"Anthropological accounts are derived from aspects of the personality of the author. How could it be otherwise? When Malinowski writes about Trobriand Islanders he is writing about himself. When Evans-Pritchard writes about the Nuer he is writing about himself."

Writing Culture and Postmodernism

The Editorial Board of *History of Anthropology* includes most of the anthropologists who participated in a seminar in New Mexico on "The Making of Ethnographic Texts", published as *Writing Culture* (1986). The book created a seismic shift in anthropology. From then on, modern anthropology was distinct from postmodern anthropology. Postmodernism rejects **grand theory** of any kind.

ANTHROPOLOGISTS ARE THUS INVITED TO REJECT THE "THEORETICAL TRUTH" AND "WHOLENESS" OF ETHNOGRAPHIC REALITY ...

OUT GOES THE POSSIBILITY OF ANY TRUE OR COMPLETE STATEMENT ABOUT CULTURE - EVEN THE POSSIBILITY OF APPROXIMATION.

MODERN ANTHROPOLOGY

POSTMODERN ANTHROPOLOGY

Leach's "temporary" fictions become the "narrative character of cultural representations" of James Clifford in *Writing Culture*. Anthropology exists *in the writing* – it is a text to be read, analysed and examined for layers of meaning like a novel, and the personality of the author should be made evident in the construction of

Postmodern Paralysis

But not everyone is happy with this state of affairs. The interpretivism of Clifford Geertz predated and prefigured the *Writing Culture* revolution in anthropology and he has become its leading exponent. British anthropologist Ernest Gellner accused Geertz of...

"encouraging a whole generation of anthropologists to parade their real or invented inner qualms and paralysis, using the invocation of the epistemological doubt and cramp as a justification of utmost obscurity and subjectivism. They agonize so much about their inability to know themselves and the Other, at any level of regress, that they no longer need to trouble too much about the Other. If everything in the world is fragmented, multiform, nothing really resembles anything else, and no one can know another (or himself), and no one can communicate, what is there to do other than express the anguish engendered by this situation in impenetrable prose?"

(*Postmodernism, Reason and Religion*, London: Routledge, 1992)

Ernest Gellner

Women in Anthropology

Is anthropology a male construct? As a professional discipline, and one that was small, marginal and considered a rather strange undertaking until well into its second half-century, this is not a simple question to answer. Proportionately, there were more women in leading positions in anthropology than any other discipline in the Western academy.

I WAS ALREADY DOING FIELDWORK WHEN THE UNIVERSITY SYSTEM IN BRITAIN WAS JUST GETTING ROUND TO ENROLLING WOMEN STUDENTS.

The same is true of Phyllis Kaberry, who was Australian by birth. Other leading names in the early generation are ...

Ruth Benedict

Ruth Bunzel

Lucy Mair

Elizabeth Colson

Audrey Richards

Monica Wilson

Hilda Kuper

Mary Douglas

Kathleen Gough

Rosemary Harris

Laura Nader

Kinship Ties of Anthropologists

Women in anthropology have been remarkable and have left their mark not just as individuals but also in two further significant ways. The first is by "power marriages" … *Margaret Mead was married to Gregory Bateson, then there's Monica and Godfrey Wilson, Hilda and Leo Kuper (uncle and aunt to Adam Kuper, spouse of Jessica Kuper); others include Robert and Elizabeth Fernea, Simon and Phoebe Ottenberg, the Goodys, the Arendts, the Marshalls, the Peltos, the Stratherns and many more.*

THESE MARRIAGES LINK TWO TRAINED WORKING ANTHROPOLOGISTS, EACH OF WHOM IS A POWER IN HIS OR HER OWN RIGHT.

WHERE THEY COLLABORATE ON FIELDWORK, THE DOMESTIC WORLD OF WOMEN AMONG THE PEOPLE STUDIED IS OPENED TO THE MALE PARTNER.

The Field Helpmate

The second impact is as the "anthropological wife". Not necessarily a trained anthropologist, but companion and helpmate of the male anthropologist in the field. If not She Who Must Be Obeyed, certainly not someone to underestimate or overlook as a silent partner, the way male anthropologists have done.

Laura Bohannan, wife of anthropologist Paul Bohannan, wrote *Return to Laughter*. Mary Smith, wife of anthropologist M.G. Smith, wrote *Baba of Karo*, the biography of a Muslim Hausa women. It gives a whole double entendre to the concept of **female agency**.

Feminist Anthropology

Edwin Ardener (1927–87), Professor of Anthropology at Oxford, suggested that anthropology itself is male dominated, not merely preponderantly male in recruitment, but that its theory, concepts, methodology and practice are artefacts of male culture even when practised by women.

Feminist anthropology is as complex and diverse as feminism. It includes questions such as Sherry Ortner's "Is Female to Male as Nature is to Culture?"

Situating Feminist Anthropology

Its most balanced proponent and exponent is Henrietta Moore. *"Feminist anthropology … formulates its theoretical questions in terms of how economics, kinship and ritual are experienced and structured through gender, rather than asking how gender is experienced and structured through culture."* Moore argues against the concept of "universal women".

Women anthropologists need be no better or worse than male anthropologists in making women's voices audible and women's agency and roles visible.

The Virgin People

The Yanomamo of the Venezuelan section of the Amazonian rainforest are the world's most famous indigenous people. They represent that most prized of all anthropological trophies – a "virgin people", allegedly untouched by white society in their remote rainforest enclave, a last vestige of how it all once was, and last of the vanishing peoples.

OH REALLY? HASN'T ANYONE HEARD OF SPANISH CONQUISTADORES, RUBBER-TAPPERS AND GOLD-MINERS IN THE AMAZON?

WE ARE ALSO THE MOST FAMOUS POSTER PEOPLE OF "SURVIVAL CAMPAIGNS", "SAVE THE RAINFOREST" AND ECOLOGICAL MOVEMENTS.

WE WERE EVEN VISITED BY AND PHOTOGRAPHED WITH POP ICON, STING.

Not surprisingly, the Yanomamo have been studied by anthropologists for decades.

The Yanomamo Scandal

The two most famous or infamous anthropologists doing fieldwork with the Yanomamo are the American Napoleon Chagnon, one of the best-known contemporary anthropologists, and Jacques Lizot, a French anthropologist who was a student of Lévi-Strauss.

Patrick Tierney, in his book *Darkness in El Dorado: How Scientists and Journalists Devastated the Amazon* (2000), presents the following case against these anthropologists.

Tierney's charges are that Chagnon ...

- Was a member of a research team funded by the American Atomic Energy Commission. The funding was for collection of Yanomamo blood samples to provide base comparison on naturally occurring background radiation. Of no earthly use to the Yanomamo and beyond possibility of informed consent, but potentially dangerous by exposing them to infection.

- Used live virus measles vaccine that caused rather than treated an epidemic of measles that decimated Yanomamo villages.

And the list of charges goes on …

- Paid Yanomamo informants in trade goods, valuable steel axes and guns in order to gather ethnographic information. Outraged Yanomamo by using this method to gather "secret" names of dead relatives. The influx of trade goods destabilized villages, causing conflict within and between villages.

- Used the trade goods to stage-manage specially rehearsed ethnographic films. Two of these films The Ax Fight and The Feast are the most famous ethnographic "documentaries".

- Showed unethical and potentially genocidal irresponsibility in taking film-makers and journalists into Yanomamo areas with little regard to their susceptibility to Western diseases.

THESE FILMS ARE THEREFORE FRAUDULENT AND MISREPRESENT THE PEOPLE THEY PRESENT TO A WESTERN AUDIENCE.

BUT THAT'S NOT ALL …

- Obfuscated research material to support untenable interpretations, especially the thesis that men who kill most have the greatest number of wives and thereby dominate the gene pool. This evidence is central in substantiating the theoretical predictions of sociobiology on how humankind originated and developed.

- Falsified how many deaths have actually occurred and who committed them, as well as invented "murders".

- Failed to provide treatment for sick and dying Yanomamo.

- Failed to provide advocacy for them.

- Attempted to establish authority over and exclusive access to the Yanomamo in collaboration with a man who was a leading profiteer in the gold-mining rush that is devastating Yanomamo lands.

- Constructed the Yanomamo as pristine people when the entire region has been under the influence and impact of colonial depredation since the arrival of the Spanish and Portuguese.

And that Lizot …

- Sexually abused Yanomamo boys to feed his own sexual fantasies.

Creating Civil War

As anthropologists, Chagnon and Lizot represent two different versions of the Yanomamo. But Tierney argues that this now familiar charge goes further. It has led to active conflict between villages that are client-dependants of each anthropologist for access to trade goods.

The anthropologists have perversely reconstructed the politics of the people they study.

The willing helpers in all this have been the Western media. Ever eager for headlines about "Stone Age" societies, they have made Chagnon rich and famous and avidly popularized his "Fierce People" thesis. The Fierce People theory recycles all the original aspects of the concept of the primitive, with the sociobiological addition that homicide is the original convention of human society, and genocide, as a logical cognate, must also be original to humanity.

[A complete defence of the total innocence of Chagnon on any and all charges made against him is to be found on the University of California, Santa Barbara Website:
http://www.anth.ucsb.edu/chagnon.html]

Whither Anthropology?

The professional history of anthropology spans barely one century. Its second half-century has been revisionist of the first. Anthropology has been a discipline in crisis, a cul-de-sac, perennially contemplating its own imminent demise.

BUT HAS DOING ANTHROPOLOGY CHANGED?

ANTHROPOLOGY IS AN ECLECTIC DISCIPLINE AND GETS MORE ECLECTIC.

Roy D'Andrade (1995), echoing Ernest Gellner, sums up how anthropology operates as "agenda hopping". Investigation reveals not only complexity, requiring more and more effort to generate anything new, but whatever is found seems less and less interesting. "When this happens, a number of practitioners may defect to another agenda – a new direction of work in which there is some hope of finding something really interesting."

Anthropology remains the "study of the Other" rather than a dialogue with the Other. Anthropology has popularized Other lifestyles that have become designer accessories of affluent Western consumerism. Eco-tourism now permits the affluent to visit the "quaint exotics" and sample for themselves what anthropology is all about. Anthropology has not assisted in equalizing power or disparities in wealth between the West and the Other, even if some anthropologists privately believe this should happen.

ALL THESE QUESTIONS AND UNCERTAINTIES WERE ALREADY FAMILIAR TO ME IN THE 16TH CENTURY.

THEY'RE ALL TOO FAMILIAR TO ME - AND THAT'S THE POINT EVERYONE STILL AVOIDS.

cover
ur resort

LULU AFRIKA SAFARIS

afaris

ne
African Desti
stinations
nture Geta

el Planner

Further Reading

Classic Ethnography and Anthropology

Boas, Franz, *The Mind of Primitive Man*, New York: Macmillan, 1938.
—— *Race, Language and Culture*, New York: Macmillan, 1940.
Evans-Pritchard, E.E., *Witchcraft, Oracles and Magic Among the Azande*,
    Oxford: Clarendon Press, 1937.
—— *Nuer Religion*, Oxford: Clarendon Press, 1956.
—— *Social Anthropology*, London: Cohen and West, 1951.
Firth, Raymond, *We, the Tikopia*, London: Allen and Unwin, 1936.
Kaberry, Phyllis, *Women of the Grassfields*, London: HM Stationery Office,
    1952.
Kroeber, A.L., *Anthropology: Culture Patterns and Processes*, New York:
    Harcourt, 1963.
Kluckhohn, Clyde, *Navaho Witchcraft*, Cambridge, MA: Peabody Museum,
    1944.
Malinowski, Bronislaw, *A Scientific Theory of Culture and Other Essays*,
    Chapel Hill: University of North Carolina Press, 1944.
—— *Argonauts of the Western Pacific*, London: Routledge, 1922.

General Introductions

Beattie, J., *Other Cultures: Aims, Methods and Achievements in Social
    Anthropology*, London: Routledge, 1964.
Bohannan, Paul, *We, the Alien: An Introduction to Cultural Anthropology*,
    Prospect Heights, IL: Waveland Press, 1992.
Geertz, Clifford, *The Interpretation of Cultures*, New York: Basic Books,
    1973.
Gluckman, Max, *Politics, Law and Ritual in Tribal Societies*, Oxford: Basil
    Blackwell, 1965.
Ingold, Tim, *Companion Encyclopedia of Anthropology: Humanity, Culture
    and Social Life*, London: Routledge, 1994.
Lewis, I.M., *Social Anthropology in Perspective*, Cambridge: Cambridge
    University Press, 1985.

History and Theory

Adams, William Y., *The Philosophical Roots of Anthropology*, Stanford: CSLI
    Publications, 1998.
Barnard, Alan, *History and Theory in Anthropology*, Cambridge: Cambridge
    University Press, 2000.
Hodgen, Margaret, *Early Anthropology of the Sixteenth and Seventeenth
    Centuries*, Philadelphia: University of Pennsylvania Press, 1964.
Kuper, Adam, *Invention of the Primitive*, London: Routledge, 1988.

Layton, Robert, *An Introduction to Theory in Anthropology*, Cambridge: Cambridge University Press, 1997.
Moore, Henrietta, *Anthropological Theory Today*, Cambridge: Polity Press, 1999.

Critique and New Directions

Deloria, Vine, Jr., *Custer Died For Your Sins: An Indian Manifesto*, New York: Macmillan, 1969.
Geertz, Clifford, *Works and Lives: The Anthropologist as Author*, Stanford: Stanford University Press, 1988.
Hymes, Dell, *Reinventing Anthropology*, revised ed., Ann Arbor: Ann Arbor Paperbacks, 1999.
Obeyesekere, G., *The Apotheosis of Captain Cook*, Princeton: Princeton University Press, 1992.
Rapport, Nigel and Overing, Joanna, *Social and Cultural Anthropology: The Key Concepts*, London: Routledge, 2000.
Sahlins, Marshall, *How "Natives" Think*, Chicago: University of Chicago Press, 1995.

About the Author and Artist

Merryl Wyn Davies is a writer and anthropologist. A former television producer who worked for BBC religious programmes, she is the author of a number of books, the highly acclaimed *Knowing One Another: Shaping an Islamic Anthropology* (1988) and co-author of *Distorted Imagination: Lessons from the Rushdie Affair* (1990) and *Barbaric Others: A Manifesto on Western Racism* (1993). Her most recent book is *Darwin and Fundamentalism* (2000). Forever Welsh, she lives and works in Merthyr Tydfil.

Piero is an illustrator and graphic designer. He earned his degree at the Art University of La Plata, Buenos Aires, and has had several published illustrations. His work has twice been included in the Royal College of Art's "The Best of British Illustration" (1998,1999). He studied animation and multimedia at Westminster College in London and illustrated *Introducing Shakespeare* (2001).

Acknowledgements

The author would like to thank Ziauddin Sardar for his ever-constructive criticism and persistence in ensuring this book got written, and Jennifer Rigby for her patient endurance.

The artist would like to thank Richard Appignanesi and dedicate this book to Mora, Rocio and Silvina.

Other Introducing Books …

In case of difficulty in purchasing any Icon title through normal channels, books can be purchased through BOOKPOST. Tel: +44 (0)1624 836000. Fax: + 44 (0)1624 837033. e-mail: bookshop@enterprise.net. Website: www.bookpost.co.uk. Please quote "Ref: Faber" when placing your order. If you require further assistance, please contact: info@iconbooks.co.uk.

Index